T0135699

Philipp Wenzelburger

# A Model Predictive Control Framework for Flexible Job Shop Scheduling

Logos Verlag Berlin

λογος

Bibliografische Information der Deutschen Nationalbibliothek

Die Deutsche Nationalbibliothek verzeichnet diese Publikation in der
Deutschen Nationalbibliografie; detaillierte bibliografische Daten sind
im Internet über http://dnb.d-nb.de abrufbar.

D93

ISBN 978-3-8325-5532-0

Logos Verlag Berlin GmbH
Georg-Knorr-Str. 4, Geb. 10,
D-12681 Berlin
Germany

Tel.:    +49 (0)30 / 42 85 10 90
Fax:    +49 (0)30 / 42 85 10 92
http://www.logos-verlag.de

*Für meine Familie*

# Acknowledgments

I would like to thank all my colleagues from the Institute for Systems Theory and Automatic Control (IST) for their support and the great time we had together. In particular I am grateful to Prof. Frank Allgöwer for his supervision during my time as a PhD student. I very much enjoyed the working atmosphere he created at the IST, which fosters the exchange among colleagues and leads to excellent research results. On a personal side, he convinced me to continue and publish my work especially in times when I had doubts myself. Although our views on the potential of my research were often quite contrary - or probably because of this - I was able to successfully complete my dissertation thanks to his continuous support.

Many important inputs to my work arose in the countless discussion we had in the MPC group. Despite my research topic is rather exotic with respect to other work on Model Predictive Control (MPC) at the IST, every group meeting was a little boost for my progress. I greatly appreciated the feedback from the members of the MPC group which revealed weaknesses of my work, suggested promising research directions and sometimes pointed out some strengths that I had overseen. Besides that, I want to thank my office mates Hans-Bernd Dürr, Simon Niederländer, Karsten Kuritz, Simon Michalowsky and David Meister for answering all the quick questions on the fly.

I also want to thank the examination committee, Prof. Meike Tilebein, Prof. Vicente Lucena and Prof. Alexander Verl for their interest in my work and the interesting although challenging examination discussion. Last but not least I am grateful to the proofreaders Janani Venkatasubramanian, Matthias Köhler (née Hirche), Henning Schlüter, David Meister, Lukas Schwenkel, Stefan Wildhagen and Raffaele Soloperto for their valuable input and for finding the smallest mistakes in my thesis.

Stuttgart, June 2022
Philipp Wenzelburger

v

# Table of Contents

# List of Acronyms and Symbols

## Acronyms

| | |
|---|---|
| FJS | flexible job shop |
| FMS | flexible manufacturing system |
| JMPM | job shop with multi-murpose machines |
| JMPMSF | job shop with multi-purpose machines and sequence flexibility |
| JS | job shop |
| MILP | mixed integer linear program |
| MIP | mixed integer program |
| MPC | model predictive control |
| PJS | partial job shop |
| PN | Petri net |

## General mathematical notation

| | |
|---|---|
| $\emptyset$ | The empty set |
| $\lceil a \rceil$ | Smallest integer value greater or equal to $a \in \mathbb{R}$ |
| $I$ | Identity matrix of appropriate dimension |
| $k$ | Time instant |
| $\mathbb{N}$ | Set of natural numbers, i.e., nonnegative integers |
| $\mathbb{N}_{>0}$ | Set of positive integers |
| $\mathbb{R}_{\geq 0}$ | Set of non-negative real numbers |
| $\mathbb{R}_{>0}$ | Set of positive real numbers |
| $t_s$ | Sampling time or sampling period, a time interval that passes between two sampling instants |
| $x_i$ | $i$-th entry in a vector $x$ |

# Scheduling notation

$\Phi$      Function to recursively determine the indirect dependencies $\bar{\mathcal{T}}_\tau = \Phi(\mathcal{T}_\tau)$ of a task $\tau$ from the set of its direct dependencies $\mathcal{T}_\tau$, cf. Algorithm 2.

$E_J$      Directed edges in the precedence graph $G_J$ of the tasks of a job $J$

$G_J$      Precedence graph of the tasks of a job $J$

$J$      A single job

$\mathcal{J}$      Set of all jobs that have to be fulfilled by the manufacturing system

$k_P(M,\tau)$      Number of production steps to execute the task $\tau$ on the machine $M$ in the discrete time model

$M$      A single machine

$\mathcal{M}$      Set of all machines available in the manufacturing system

$O$      A single operation, i.e., a pair $O = (\tau, J)$ of a task $\tau$ and a job $J$

$\mathcal{O}$      Set of all operations, i.e., the set of all pairs $O = (\tau, J)$ with $\tau \in \mathcal{T}_J$, $J \in \mathcal{J}$

$\tau$      A single task

$\mathcal{T}$      Set of all tasks executable in the manufacturing system

$\mathcal{T}_M$      Set of all tasks that a machine $M$ can execute, cf. restriction (R5)

$\mathcal{T}_J$      Set of all tasks that need to be completed in order to finish a job $J$

$\mathcal{T}_\tau$      Set of tasks that have to be completed before the task $\tau$ can be started, cf. Restriction (R1). Only *direct* dependencies are captured in $\mathcal{T}_\tau$.

$\bar{\mathcal{T}}_\tau$      Set of all tasks that have to be completed before the task $\tau$ can be started. In contrast to $\mathcal{T}_\tau$, $\bar{\mathcal{T}}_\tau$ it reflects direct *and* indirect dependencies, meaning that it also contains all tasks $\tau' \in \bar{\mathcal{T}}_\tau$ and $\tau'' \in \bar{\mathcal{T}}_\tau$ with $\tau'' \in \bar{\mathcal{T}}_{\tau'}$ or $\tau' \in \bar{\mathcal{T}}_{\tau''}$.

$t_P(M,\tau)$      Production time of task $\tau$ on machine $M$

$\bar{t}_P(M,\tau)$      Production time of a task $\tau$ on a machine $M$ in the discrete time model, i.e., $\bar{t}_P = k_P t_s$

# Petri net notation

$A$      Dynamics matrix of a linear time invariant system, in this thesis most of the time representing the independent part of a Petri net considering the independent transitions $T \in \mathbb{T}_I$

$B$      Input matrix of a linear time invariant system, in this thesis most of the time representing the incidence matrix of a Petri net considering the controlled transitions $T \in \mathbb{T}_C$

$B^-$      In-Incidence matrix of the transitions of a Petri net representing the influence of the arcs $(P, T) \in \mathbb{E}$

$B^+$      Out-Incidence matrix of the transitions of a Petri net representing the influence of the arcs $(T, P) \in \mathbb{E}$

| | |
|---|---|
| $\mathbb{E}$ | Set of arcs in a Petri net |
| $m$ | Number of transitions in a Petri net |
| $n$ | Number of places in a Petri net |
| $P$ | A single place in a Petri net |
| $\mathbb{P}$ | Set of all places in a Petri net |
| $PN$ | Petri net |
| $\sigma$ | Classifier for the production related meaning of a place or transition in the Petri net |
| $\Sigma$ | Set of classifiers for the production related meaning of the places and transitions in the Petri net; $\Sigma = \{S, F, P_q, B, I, N, C\}$ |
| $T$ | A single transition in a Petri net |
| $\mathbb{T}$ | Set of all transitions in a Petri net |
| $\mathbb{T}_{O,S}$ | Set of transitions that start the execution of operation $O$ |
| $\mathbb{T}_C$ | Set of controlled transitions in a Petri net |
| $\mathbb{T}_I$ | Set of independent transitions in a Petri net |
| $u$ | Firing count vector in a Petri net and input vector to the manufacturing system |
| $u^{\mathrm{ID}}$ | Vector of identifiers corresponding to the inputs in a Petri net model. It holds the identifiers of the controlled transitions. |
| $u_T$ | Input vector which only fires a single starting transition |
| $\mathbb{U}$ | Set of inputs $u$ |
| $\mathbb{U}_{O,S}$ | Set of input vectors that start the execution of operation $O$ |
| $w$ | Arc weight function $w : \mathbb{E} \to \mathbb{N}_{>0}$, i.e., a function assigning a weight to each arc in a Petri net |
| $x$ | State of a Petri net or a dynamical system |
| $x^0$ | Initial state of a Petri net |
| $x^{\mathrm{ID}}$ | Vector of identifiers corresponding to the states of the Petri net model. It holds the identifiers of the places. |
| $\mathbb{X}$ | Set of states $x$ |
| $\mathbb{X}_{O,S}$ | Set of all states in which the operation $O$ can be started, cf. Definition 3 |
| $\mathbb{X}_{O,C}$ | Set of all states right after completion of operation $O$, cf. Definition 4 |

## MPC notation

| | |
|---|---|
| $c$ | Cost function $c : \mathbb{X} \times \mathbb{U} \to \mathbb{R}$ that quantifies the cost induced by the state $x$ of a dynamical system and the input $u$ applied to it. It is called *stage cost* in MPC literature. |
| $\bar{k}$ | Future time instant in the MPC problem, where counting is started at time instant $k$. The MPC predicts the behavior of the system for a time step $\bar{k}$ in the future, which is the time $k + \bar{k}$ counting from initialization of the system. |

$N$ Prediction horizon in the MPC Problem

$N_O$ Sufficiently long prediction horizon for the MPC problem (4.2) that guarantees the completion of the operation $O$

$N_J$ Sufficiently long prediction horizon for the MPC problem (4.2) that guarantees the completion of the job $J$

$N_{\min}$ Sufficiently long prediction horizon for the MPC problem (4.2) that guarantees the completion of a given production problem

$N^+$ Extended prediction horizon that captures the effects after the prediction horizon $N$. It is used in the MPC with terminal cost (4.12). During the extended prediction horizon, no optimization over the possible input trajectories is carried out, but a cost value of the terminal state is estimated.

$N_O^+$ Sufficiently long extended prediction horizon for the MPC problem with terminal cost (4.12) that guarantees the completion of the operation $O$

$N_J^+$ Sufficiently long extended prediction horizon for the MPC problem with terminal cost (4.12) that guarantees the completion of the job $J$

$N_{\min}^+$ Sufficiently long extended prediction horizon for the MPC problem with terminal cost (4.12) that guarantees the completion of a given production problem

$u(\cdot|k)$ Predicted input trajectory in the MPC problem initialized at time instant $k$. The symbol "$\cdot$" takes arbitrary values for the future time instants $\bar{k}$.

$u(\bar{k}|k)$ Predicted input to the system at the future time instant $k + \bar{k}$ in the prediction in the MPC problem initialized at time instant $k$

$u^*(\cdot|k)$ Minimizer of the MPC problem, i.e., the input trajectory $u(\cdot|k)$ which leads to the minimum cost in the prediction

$V_{\mathsf{f}}$ Terminal cost function $V_{\mathsf{f}} : \mathbb{X} \to \mathbb{R}$ that quantifies the cost of a terminal state $x(N|k)$ in an MPC problem

$x(\cdot|k)$ Predicted state trajectory in the MPC problem initialized at time instant $k$. The symbol "$\cdot$" takes arbitrary values for the future time instants $\bar{k}$.

$x(\bar{k}|k)$ Predicted state of the system at the future time instant $k + \bar{k}$ in the prediction in the MPC problem initialized at time instant $k$

$x^*(\cdot|k)$ Predicted state trajectory resulting from applying the minimizer $u^*(\cdot|k)$ of the MPC problem to the model starting from the measured state $x(0|k) = x(k)$

$\mathbb{X}_{\mathsf{f}}$ Set of final states in the MPC problem

# Abstract

In this thesis, we develop a framework for the reactive scheduling of flexible manufacturing systems based on model predictive control (MPC). We integrate the modeling and scheduling of a flexible manufacturing system in a common framework, in which the flexibility of the system can be exploited, the modeling effort is minor, and guarantees for the resulting behavior of the manufacturing system are provided.

The framework is specifically designed for manufacturing systems intended for the production of customer specific goods, which is motivated by the concept of Industry 4.0. In this context, the machines are assumed to have a known set of capabilities and the products have a known set of required production steps. Based on a modular description of the manufacturing system, which comprises multiple machines and which is intended to produce a variety of customer specific goods, the production steps have to be assigned to the machines in the best possible way respecting their capabilities. This modular production system offers flexibility, which can only be exploited by a suitably designed scheduling framework as the one proposed in this thesis.

The basis of this framework is a description of the manufacturing system in the form of a flexible job shop. In order to cover a wide range of possible application cases and introduce flexibility from the start, the problem description is kept general and modular. Therefore, there is the possibility to deduce numerous widely discussed scheduling problems as special cases of the proposed problem description, all of which can also be considered in the developed framework.

From the modular description of the flexible job shop, a Petri net model is automatically generated by means of specifically designed algorithms. The dependencies between the different parts of the problem description are captured in the Petri net graph, while it preserves the flexibility of the problem formulation. This is achieved by representing the possible decisions in the manufacturing system through the different possibilities to fire the transitions in the Petri net. In accordance with that, the marking of the Petri net represents the status of the manufacturing system and restricts the possible decisions according to the dependencies in the production problem. Moreover, the algebraic description of the Petri net is well suited for a system theoretic analysis and the application of MPC.

Through the automatic conversion of the production problem, the MPC scheme developed on the basis of the Petri net model retains the properties of the original problem formulation. Incorporating the economic objective of the manufacturing system into the cost function employed in the optimization problem solved in the MPC scheme allows to pursue the economic goal in the MPC based scheduling framework. Through analyzing the properties of the manufacturing system and its algebraic representation, we are able to prove the recursive feasibility of the MPC problem and the completion of the production problem through the proposed scheduling scheme. By considering feedback from the system, the MPC is able to react to unexpected events and changes in the manufacturing system.

The proposed framework is able to make optimal use of the manufacturing units in a flexible manufacturing system for the production of customer specific goods. The advantages of the framework are its applicability to a wide range of scheduling problems, its ability to model and to exploit the available flexibility in the manufacturing system for economically optimal operation, and its guarantee to complete the production problem.

# Deutsche Kurzzusammenfassung

**Ein Framework für die modellprädiktive Regelung in der flexiblen Produktionsplanung**

In dieser Arbeit wird ein Framework für die reaktive Produktionsplanung von flexiblen Fertigungssystemen mittels modellprädiktiver Regelung (Englisch: *model predictive control*, MPC) entwickelt. Wir vereinen dabei die Modellierung eines flexiblen Fertigungssystems mit der Produktionsplanung und -regelung in einem gemeinsamen Framework, in dem die Flexibilität des Systems ausgenutzt werden kann, der Modellierungsaufwand gering ist und systemtheoretische Garantien für das resultierende Verhalten des Fertigungssystems gegeben werden können.

Motiviert durch die Entwicklungen in der Industrie 4.0 wird das Framework speziell für Fertigungssysteme entwickelt, die für die Herstellung kundenspezifischer Produkte konzipiert sind. In diesem Zusammenhang wird davon ausgegangen, dass die Fähigkeiten der Maschinen und die für die Herstellung der Produkte erforderlichen Produktionsschritte bekannt sind. Ausgangspunkt ist eine modulare Beschreibung des Fertigungssystems, das aus mehreren Maschinen besteht und in dem eine Vielzahl von kundenspezifischen Gütern hergestellt werden sollen. In einem so konzipierten Produktionssystem müssen die zur Herstellung der Produkte erforderlichen Produktionsschritte jenen Maschinen zugeordnet werden, die sie aufgrund ihrer Fähigkeiten unter den gegebenen Rahmenbedingungen am effizientesten produzieren können. Dieser modulare Aufbau des Produktionssystems bietet eine große Flexibilität, die es durch eine geeignete Produktionsplanung auszunutzen gilt, was das in dieser Arbeit entwickelte Framework ermöglicht.

Ausgangspunkt des vorgestellten Frameworks ist eine Beschreibung des Produktionssystems in Form eines flexiblen Job-Shops. Um ein breites Spektrum an möglichen Anwendungsfällen abzudecken und schon bei der Problembeschreibung die Flexibilität zu berücksichtigen, wird das Problem sehr allgemein und modular formuliert. Die Problembeschreibung umfasst daher zahlreiche in der Literatur betrachtete Produktionsplanungsprobleme als Spezialfälle und ermöglicht es, diese im entwickelten Framework ebenfalls zu betrachten.

Aus der modularen Beschreibung des flexiblen Job-Shops wird mittels speziell entwickelter Algorithmen automatisch ein Petri-Netz-Modell generiert. Dieses behält die Flexibilität der Problemformulierung bei und in seiner Graphenstruktur sind die Restriktionen und Abhängigkeiten berücksichtigt, die das Produktionsplanungsproblem beinhaltet. Die Markierung des Petri-Netzes repräsentiert den Zustand des Produktionssystems und die Möglichkeiten, wie die Transitionen schalten können, stellen die möglichen Entscheidungen im Produktionssystem dar. Die algebraische Beschreibung des Petri-Netzes ist gut für eine systemtheoretische Analyse und die Anwendung von MPC geeignet.

Bei der automatischen Transformation des Produktionsproblems in die MPC-Formulierung werden die Eigenschaften der ursprünglichen Problemformulierung beibehalten. Durch das Einbeziehen des ökonomischen Ziels des Produktionssystems in die Kostenfunktion, die bei der Optimierung im MPC-Algorithmus minimiert wird, wird das ökonomische Ziel bei der MPC-basierten Produktionsplanung explizit berücksichtigt. Auf Basis der Analyse der Eigenschaften des Produktionssystems und seiner algebraischen Darstellung kann garantiert werden, dass das MPC-Problem, welches dem entwickelten Produktionsplanungsverfahren zu Grunde liegt, zu jeder Zeit lösbar ist und dass das Produktionsproblem durch den MPC-Ansatz fertig gestellt wird. Durch die Berücksichtigung von Informationen, die aus dem System zurückgeführt werden, kann die MPC-basierte Produktionsplanung auf unerwartete Ereignisse und Veränderungen im Produktionssystem zielgerichtet reagieren.

Das vorgestellte Framework zur Produktionsplanung ist in der Lage, die Produktionseinheiten in einem flexiblen Fertigungssystem, in dem kundenspezifische Aufträge erfüllt werden, in der bestmöglichen Art und Weise einzusetzen. Die Vorteile der vorgeschlagenen Methode sind, dass sie für eine Vielzahl verschiedenartiger Produktionsplanungsprobleme einsetzbar ist, dass sie in der Lage ist, die verfügbare Flexibilität zu modellieren und im Produktionssystem auszunutzen, um einen wirtschaftlich optimalen Betrieb der Anlage anzustreben, und dass sie das Produktionsproblem garantiert fertigstellt.

# Chapter 1

# Introduction

## 1.1 Motivation

Industrial production systems are facing the need to stay competitive in a changing world. In the modern society, where the basic needs of the customer are increasingly satisfied, innovation cycles are getting shorter and the products are adjusted to the specific needs of individual customers. In order to account for this development, the fourth industrial revolution has been proclaimed in Germany and is now being implemented by industry worldwide. This endeavor is called *Industry 4.0* and it exploits emerging new technologies, mainly from the field of information and communication technology, with the goal of developing smart factories that account for customers' individual wishes and generating new business models [39]. Especially networking and digital communication capabilities are employed in manufacturing units and are considered as drivers of this development and expected to revolutionize industry. Smart factories with ubiquitous communication between all components are envisioned to produce smart products that know by themselves how they have to be produced [39].

This environment of flexible interconnections between the participants requires novel strategies for the scheduling of the increasingly dynamic manufacturing system. The production scheduling is challenged with the goal of enabling the production of highly individual goods at the same cost as mass production in order to stay competitive [14]. This challenge is in direct continuation of the history of production scheduling, which started at the end of the 19th century, when the product variety in the individual manufacturing facilities increased [31]. The increasing complexity of the production system already at that time led to the organized production planning based on the ideas of Frederick W. Taylor and Henry L. Gantt. Dedicated planning offices were introduced and the production schedules were represented in the form of Gantt charts, which are still widely used. The underlying production problems can be formulated as optimization problems and the invention of electronic computers in the second half of the 20th century facilitated their solution through the automatic execution of complex optimization algorithms. By that, the time to create production schedules and to optimize the production processes could be reduced. Due to the complexity of the scheduling problems, mainly techniques to find suboptimal solutions in a short amount of time were developed [24]. Those include various heuristics and in modern

1

days also concepts using machine learning techniques attract increasing attention [13, 85]. For exact solutions of scheduling problems, mathematical programming based on mixed integer program (MIP) formulations of the scheduling problem are used, but they are limited to small and medium sized problems [15, 24]. They suffer from the high complexity of the underlying problems and the resulting large computation times and sometimes even infeasibility for the case of large problem instances, but have the advantage that they offer the possibility to provide rigorous guarantees. The more detailed review of the history of production scheduling by Herrmann [31] reveals that since the advent of information technology, the development of novel scheduling techniques was always driven by the increase in computational power, making previously intractable problems solvable.

Also in the current development of Industry 4.0, the increase in communication and computation capabilities offers the basis for the development of novel scheduling techniques. The new capabilities foster the development of modular and flexible production paradigms such as skill-based engineering, cyber-physical production systems and digital twins of all components involved in the production system [2, 61, 70]. They exploit the miniaturization of computational units combined with communication interfaces to form self-aware components that interact with each other. Such modular units need to be orchestrated by advanced scheduling techniques.

Although there is no particular joint development of Industry 4.0 and scheduling research, both developments go in a similar direction [70, 84]. The increasing complexity of the manufacturing environment motivates both the development of new concepts in Industry 4.0 and the further development of scheduling techniques. For the two lines of research to mutually benefit, a unified framework for data exchange and execution of manufacturing tasks is needed to allow the implementation of scheduling algorithms in Industry 4.0 environments [84].

The general goal of industrial manufacturing is to deliver the highest economic outcome while consuming the least possible resources and thereby optimize the manufacturing process [39]. In a dynamic manufacturing environment as it is envisioned in Industry 4.0, this requires optimization of the manufacturing system and in particular of the production schedules. The production of individual goods needs to be achieved by the highly flexible machines in an optimized fashion in order to be competitive. The products are aware of their required production steps and the machines provide their capabilities. Both are available in a digital representation of the production system by means of their digital twins [61]. In the digital environment, the optimization of the production can take place and the optimal production scheduled can be computed in real-time based on the most up-to-date information about the manufacturing system. The optimization goal, although it usually considers an economic objective by trying to maximize the profit generated with a given production system, might also consider further criteria as the reduction of waste or the environmental impact [69]. Through repeated optimization of the production schedule, the flexibility of the manufacturing system can be exploited by reacting to the request for new products to be scheduled and unexpected changes in the production system and its surrounding.

In order to support this development, we exploit the strengths of mathematical programming based scheduling techniques to provide mathematically rigorous guarantees for the optimization of the production schedule in real-time and propose a model predictive control (MPC) based approach. MPC is a modern control technique that has a wide variety of applications ranging from complex plants in process industry to trajectory tracking for autonomous vehicles [22, 50]. It heavily relies on a mathematical model of the system to be controlled. Based on this model and a measurement of the most recent information from the system, a finite horizon optimal control problem is solved and only the first part of the optimal solution is applied as an input to the system. This procedure of measuring the state, solving the optimal control problem and applying the input is repeated at every time instant in a closed feedback loop [67]. The models being used can be linear, nonlinear, time or parameter varying or even distributed among multiple controlled agents [28, 52, 67]. In an MPC approach to production scheduling as proposed in this thesis, based on a system model the behavior of the production system is predicted and optimized over a finite time horizon into the future. From the resulting optimal decisions in the planned production process, only the decisions for the current time instant are applied to the system. Through measurements in the production system and information on new jobs that have to be scheduled, feedback is introduced and the optimization is repeated in the next time step based on the new information. The advantages of MPC are, besides the sound understanding due to the wide range of research and applications, that it solves an underlying optimization problem and therefore tends towards an optimal system behavior with respect to a desired performance criterion and that it is able to explicitly consider state and input constraints. Another advantage, which is obvious when it is applied to classical control problems but is relevant with respect to the use of MPC as scheduling technique, is the fact that it has an inherent feedback mechanism while optimizing the system behavior. By repeatedly solving a finite horizon optimization problem it is computationally efficient with respect to solving the entire scheduling problem at once.

In this thesis, we propose a framework in which the advantages of MPC are exploited to incorporate instanteneous feedback from the production process into the digital decision making on the scheduling of the plant and optimize its future behavior. We contribute with control theoretic methods toward a real-time optimal production in flexible manufacturing systems (FMSs) intended to fulfill customer specific orders motivated by Industry 4.0. In the proposed framework, we combine the MPC based scheduling scheme with automatic model generation for production systems. Thereby, not only changes in the state of the system, but also in the system structure can be considered during runtime. We exploit the modularity of the production system as basis for the formulation of the model generation algorithms, which can also incorporate changes of the system at a later point in time into the system model. The model generation, which is based on a formulation of the manufacturing system in the form of a flexible job shop (FJS), results in a Petri net (PN) model, which stands out for its simplicity and the resulting linear state space model, which can be used in MPC.

3

To summarize, the goal of this thesis is the formulation of a framework for the scheduling of a manufacturing system, which exploits the flexibility of the production system and provides theoretical guarantees based on a mathematical model. Through the interdisciplinary research, we combine the strengths of multiple research fields to achieve an improved overall scheduling system. The modularity of the manufacturing system modeled as an FJS is exploited in the automatic model generation. The resulting system model in the form of a PN is directly suitable for MPC. Trough the inclusion of an economic objective in the computation of the MPC control law used as the scheduling scheme, the economic aspect of the manufacturing system is considered. The result of the scheduling scheme are manufacturing decisions, which are directly applied to the system. The flexibility of the manufacturing system is respected in the flexibility of the framework through the model generation and the feedback interconnection. A more detailed description of the contributions of this thesis is provided after describing related literature from the adjacent research fields.

## 1.2 Research Topic Overview

The research presented in this thesis spans multiple disciplines, starting from its motivation in Industry 4.0, over the problem formulation as a scheduling problem and the modeling in the form of a Petri net, until the solution approach by means of model predictive control. Therefore, to outline relevant existing results for this interdisciplinary research, we briefly present related literature in the respective areas with a focus on the intersection between the different fields. Parts of this research overview are based on [88] and are adopted in parts literally.

**Industry 4.0**

The development of Industry 4.0 started in Germany with the attempt to improve the competitiveness of German industry. To this end, in 2011 a working group was initiated that identified relevant research directions and outlined concepts for the future of industrial production [39]. They did not only focus on the industrial production itself, but also on adjacent topics as, for example, business and society, that are in strong relation to the industrial developments. In their final report, they deduced recommended actions to stay competitive in the future industrial environment which have been further developed since then. One of the key factors they identified is the flexibility and adaptability of the manufacturing environment to meet changing customer demands and volatility in markets [84]. To coordinate future research activities, the "Plattform Industrie 4.0" was initiated. From this starting point, research was conducted in various directions and new initiatives with similar scope were started in many countries, for example the Industry IoT Consortium (formerly Industrial Internet Consortium) in the United States, which joins forces with the Plattform Industrie 4.0 to coordinate their developments on specific topics, e.g., on the concept of the digital twin [8].

Research results on the different topics in Industry 4.0 from theory and practice are summarized in [34], which is an online handbook that is continuously updated over time and new printed editions will be released repeatedly.

In the context of this thesis, especially the research and developments on manufacturing systems with regard to Industry 4.0 are of interest. The development of intelligent manufacturing systems is motivated by the requirement of improved productivity and faster adaptation to the demand of customers [93]. In a review on different approaches and technologies towards intelligent manufacturing systems in Industry 4.0, Zhong et al. [93] identified flexibility of the manufacturing systems as one of their key features required to achieve this endeavor. They discuss multiple technologies that allow to increase the flexibility of a manufacturing system, as for example service-oriented architectures and radio-frequency identification (RFID). One way of achieving flexibility is the introduction of service-oriented or skill-based engineering [54, 62], leading to a modular orchestration of the manufacturing units enabling the coordination of autonomous agents in Industry 4.0 scenarios. Malakuti et al. [54] formulate a skill-based engineering model with the goal to enable mass customization, meaning that customized products can be created at the same cost as in mass production. Based on this model, further research question are posed on the coordination in manufacturing systems, which can be formulated as a modular scheduling problem between autonomous product agents and autonomous machines. In a similar approach to robot programming, the commands to a robot or machine are commanded in an abstract way, such that the linking between different production steps can be performed on a high level [29, 79]. This became popular in the field of robot programming, where the high level of abstraction is provided to the user in order to simplify the usage of versatile robots [2, 30, 32, 47, 71, 83].

To summarize, Industry 4.0 is a recent research initiative to which we contribute with methods from systems and control theory in order to exploit the increasing flexibility of manufacturing systems with the goal of enabling mass customization.

**Production Scheduling**

Scheduling problems arise in various different fields, e.g., the scheduling of takeoffs and landings on airports, the scheduling of games in sports tournaments or the scheduling of computations on a computer processor [35]. Various methods to solve a wide variety of scheduling problems are well studied and fill many text books, for example [10, 66]. In our research, we focus on the scheduling in industrial production and in particular on the scheduling in modular and flexible manufacturing systems that are intended for the production of individual goods and able to react to disruptive events, as for example machine breakdowns. This is in contrast to other branches of production scheduling, where the same products are repeatedly scheduled on the same machines, as it is usually the case in chemical production scheduling [76]. For the representation of scheduling problems for FMSs, formulations as job shop (JS) problems and generalized versions of it are used [6, 70], which we formally introduce in Section 2.2.

The class of problems considered in this thesis is a further generalization of the FJS, which itself is a generalization of the JS. We classify the considered problem class as job shop with multi-purpose machines and sequence flexibility (JMPMSF) in Section 2.2 according to a classification scheme introduced by Graham et al. [26]. For the solution of scheduling problems for the FJS mainly heuristics are used [13, 63], and also the even more complex JMPMSF, is mostly solved by means of heuristic approaches [7, 40, 82]. However, since we are not only interested in finding a solution for the problem but also to rigorously analyze it, we develop a mathematical modeling technique of the scheduling problem. Therefore, we do not review the heuristic approaches in more detail.

As mathematical models of the FJS, different MIP models have been developed, which were evaluated by Demir and İşleyen [15] with respect to their complexity and computational efficiency when minimizing the total makespan of the FJS. With respect to a mathematical model of the JMPMSF, Birgin et al. [6] and Özgüven, Özbakır, and Yavuz [63] present two different mixed integer linear program (MILP) models to find a schedule that minimizes the makespan.

Özgüven, Özbakır, and Yavuz [63] extended a well-known MILP model by Manne [55] for the minimization of the makespan which is computationally effective compared with other MILP models of the JS problem [15]. Through two extensions they augment the mathematical model from the JS to the JMPMSF. A simulation study confirms the applicability of the proposed model.

Birgin et al. [6] propose another MILP model for the JMPMSF based on a notion of the problem description as a graph. The problem description generalizes the one considered by Özgüven, Özbakır, and Yavuz [63] by allowing so called "Y-jobs", in which two intermediate products, which are started separately, are combined at a later stage. In a comparison with the model presented in [63], Birgin et al. [6] conclude that their model has less optimization variables and performs better most of the times in an extensive simulation study.

The practical relevance of the JMPMSF was shown by Lunardi et al. [51], who consider a JMPMSF from online printing industry that is subject to additional challenges. They present a MILP and a constraint programming model for it, after providing a concise literature review on the JMPMSF.

To react to changes and disturbances in the production system, reactive scheduling techniques are used. There are two basic approaches to react in those cases, which have been investigated by Fahmy, ElMekkawy, and Balakrishnan [18] for the case of FJS scheduling. On the one hand, the existing schedule can be taken as a starting point and it is adjusted to incorporate the required changes, for example by inserting new jobs in an existing schedule. On the other hand, changes in the system can trigger a restart of the scheduling completely from scratch. Kopanos et al. [43] investigated the influence of rescheduling including the effect induced by the change of the schedule. They include rescheduling penalties to decrease the frequency of rescheduling actions, mitigating their negative effects and making the rescheduling scheme more suited for scheduling problems in practice. Research on rescheduling and reactive scheduling in industrial production continues to these days and still gives rise to novel results. In a recent paper, McAllister, Rawlings, and Maravelias [56] analyze the effects of rescheduling and propose an improved MPC scheme for chemical production scheduling.

Motivated by the characteristics of Industry 4.0, Rossit, Tohmé, and Frutos [70] introduced a novel alternative to those general approaches for production rescheduling. They propose a smart scheduling system, which is able to react to disruptive changes in the manufacturing system and uses *Tolerance Scheduling* to generate production schedules in a JS environment. At first, a nominal schedule is computed and only events changing the quality of this schedule significantly will trigger a rescheduling. To determine when rescheduling is necessary, deviations of the delivery dates of the different products are determined, which still lead to an acceptable level of suboptimality with respect to the original schedule. Only if the real process deviates more than those values, rescheduling is initiated.

This brief review on scheduling problems is by no means exhaustive. We only considered the most relevant papers on the scheduling problems arising in Industry 4.0 and in particular on the FJS and the JMPMSF. As models for those problems, we specifically considered mathematical descriptions and thereby excluded the majority of literature, which solves the FJS and the JMPMSF by means of heuristics. To draw a brief conclusion, despite scheduling being a mature research field, there are still new problems to be considered and novel approaches are developed for their solution. A significant challenge in production scheduling is the interaction between the scheduling scheme and the real system. The disruptive effects in the production system together with changing external conditions in Industry 4.0 pose a major challenge for the scheduling schemes applied in future manufacturing systems.

To summarize, existing literature mainly covers specific scheduling problems and most frequently they are solved by means of heuristics. In this thesis, we propose a modeling and production scheduling framework that can be applied to the JMPMSF, which is a relevant class of scheduling problems in practice and generalizes further scheduling problems, in particular the frequently considered JS. The proposed framework and allows to give rigorous guarantees of the closed loop interconnection with the real system.

**Petri nets**

Petri nets (PNs) have been introduced by Carl Adam Petri in his Ph.D dissertation on the communication in automata [64]. He defines PNs as graphs through which discrete markers called *tokens* are moved according to specific rules, as we explain in more detail in Section 2.3. The main application of PNs is in the description of discrete-event systems and therefore they are an integral part of many books on this topic, e.g., [11, 73]. Starting from the initial description by Petri [64], many formalisms extending the original description have been introduced, for example timed PNs, high-level PNs and stochastic PNs [59]. The appeals of PNs are their compact representation of systems with a large state space, their modularity and that they have a graphical as well as a mathematical representation [73]. Especially their modularity makes them well suited as a modeling technique in this thesis.

PNs have been applied to model FMSs in an Industry 4.0 context. Long, Zeiler, and Bertsche [49] use extended colored stochastic PNs to model the flexibility of a production system intended for the mass production of customer specific products. They extend the classical notion of PNs by further elements in order to represent the production system and its flexibility. The resulting model can be used to support the decisions on batch sizes, product variants, delivery time and prices of the products. Latorre-Biel et al. [45] propose an object PN model as decision making support for FMSs in the frame of mass customization and Industry 4.0. They employ complex tokens, which are themselves modeled as PNs, to represent the products. Those product tokens are moved through an overarching PN representing the FMS. The two kinds of PNs are synchronized by requests for services from product agents to the machines. The modular composition of the two types of PNs jointly represents the production system. Those two publications are only exemplary to show how PN models are used to represent Industry 4.0 scenarios. They particularly underline the importance of the modularity and variability of PNs, which help to model the proclaimed flexibility of the manufacturing systems.

Yadav and Jayswal [90] provide a review of several modeling techniques for FMSs which involves, among others, PN models. They show that PN models have a wide variety of applications and are employed for deadlock avoidance, minimization of the makespan, performance analysis, scheduling and control of the FMS. They highlight that the usefulness of the PN model lies in the well-structured nature of the relationships between the modules in the PN and the simplicity of the resulting linear state space description. PN descriptions for scheduling problems of FMSs usually either directly use a real-valued production time in a timed Petri net [73], or ignore the time information and describe the production process as a discrete-event system, which again results in a non-timed PN as introduced by Petri [64].

As a conclusion, Petri nets proved to be a versatile modeling tool over the last decades. In addition to their original definition, they offer a wide variety of extensions that increase their expressive power at the expense of their clear and simple representation. Applications for the decision making process in FMSs and the modeling of FJSs indicate their usefulness in an Industry 4.0 context, for which they are particularly suited due to their modularity. In accordance with those insights from literature, we exploit the useful properties of PNs by employing them as modeling tool in the developed framework for the scheduling of FJSs.

**Model Predictive Control**

Model predictive control has been established as a versatile control scheme, in which an open loop optimal control problem based on a system model is repeatedly solved in a feedback interconnection with the real system. It is well studied, has a wide range of applications in various industrial fields [22], and several textbooks describe the results in different branches of MPC theory [28, 67]. The existing results form a mature theoretical basis and cover a wide range of different system classes. Most results on MPC consider the cases of setpoint stabilization or trajectory tracking, in which the cost function employed in the open loop optimal control problem is designed in a particular way that leads to the desired stability or tracking guarantees. In contrast to that, beyond the stabilization problem, the economic objective of the industrial plant would be the natural choice for the cost function. Investigation on the direct applications of such economic cost functions in MPC problems are considered in the theory on economic MPC. For this branch of MPC research, there exist various different approaches that are summarized in a survey by Faulwasser, Grüne, and Müller [19]. In this thesis, we employ MPC for the particular goal of reactive scheduling of FMSs modeled as PNs. To this end, we focus on the interface between MPC, production scheduling and PNs in the remainder of this literature overview.

In scheduling problems in manufacturing systems, discrete decisions to influence the future evolution of the system have to be taken. This results in discrete valued systems, for which MPC is not applied very often due to the computational complexity of the underlying optimization problem [12]. In order to simplify the optimization, Zhang, Liu, and Pannek [92] relaxed the integer constraint that arises in a dynamic capacity adjustment problem for a production system with reconfigurable machine tools which exhibits JS characteristics. Based on the relaxed formulation of the optimization problem, they optimize the capacity adjustment by means of MPC. The question how the underlying integer assignment problem can be solved based on the solution of the relaxed problem is left as an outlook. An MPC approach for the scheduling of reentrant manufacturing lines motivated by semiconductor manufacturing was considered by Vargas-Villamil and Rivera [81], who optimize the long-term behavior of the system modeled as a discrete time flow model with respect to a multi-objective cost function by means of linear programming. The result of the MPC solving the linear program is fed to a subordinate controller which computes the integer valued short-term decisions and applies them as control input to the plant. Cataldo, Perizzato, and Scattolini [12] use MPC for the scheduling in a machine environment with parallel identical machines that are fed through transportation lines of different lengths. Their goal is to maximize the production output while limiting the energy consumption. The resulting optimization problem is a MIP assigning speeds to the machines and binary inputs to the transportation system. The three publications [12, 81, 92] show at exemplary cases how MPC can be applied to discrete decision processes. They illustrate that the discrete decisions can either be relaxed and decided in an underlying problem, or directly included in the optimization problem.

In contrast to problems in scheduling and discrete manufacturing where MPC is rarely used [12], it has a variety of applications in supply chain management and especially in process industry, where continuous models naturally arise [19, 41, 76, 77]. Due to the large body of knowledge on various properties and different implementations of MPC schemes, it became a well-established control method in this area [22]. In literature which is more closely related to the topic of this thesis, MPC is applied for the scheduling of a chemical production systems in which binary decision variables are used [56, 76]. In the considered MPC problems, the possible tasks are assigned to the available production units in order to satisfy an external demand or to optimize the economic performance of the system. The problems considered in this line of research investigate the scheduling in chemical batch production systems with a previously known set of products.

In the work by Subramanian, Maravelias, and Rawlings [76], a discrete time state space model of a chemical production system is proposed. Based on this model, they provide an MPC formulation of the scheduling and batch sizing problem for the production system. It is shown how typical scheduling disruptions and further characteristics of scheduling problems, as for example resource handling, can be included in the state space model. They suggest to consider scheduling problems as economic MPC problems and to develop new scheduling algorithms that include classical notions from MPC theory, in particular recursive feasibility, closed loop performance and stability. In a recent publication, McAllister, Rawlings, and Maravelias [56] built upon this result and investigated production rescheduling with the goal of reducing frequent changes in the planned schedules by means of rescheduling penalties included in the cost function employed in the economic MPC problem. They have a particular focus on retaining theoretical properties despite the penalties on the rescheduling actions. The provided results rely on suboptimal MPC theory, in which the optimal solution does not necessarily need to be found in order to achieve the desired guarantees. McAllister, Rawlings, and Maravelias [56] are able to guarantee a certain level of performance with respect to a known reference schedule. This reference schedule is used as terminal equality constraint in the MPC problem to ensure recursive feasibility and closed loop performance with respect to the given cost function. Since the MPC inherits the performance of the reference schedule as worst case performance bound, a good reference schedule created by an elaborate heuristics is guaranteed to lead to good performance that can only be improved by the MPC.

Using systems and control theory to analyze and influence PNs is common in the literature [3, 25], and also MPC techniques are used [38, 53, 78]. In the context of MPC for PNs, usually either continuous or hybrid PNs are considered from the start, or *fluidification* of the PN is used. Fluidification means that the tokens, which in an ordinary PN are discrete markers that are moved through the PN, are considered to be a fluid. Thereby they can be represented by real numbers instead of integers relaxing the integer constraint usually imposed on the PN. Instead of a discrete state vector and discrete inputs to the system, a continuous state vector is considered and the input continuously controls the flow of tokens through the PN. This flow can usually only be in one direction and is limited by a maximum flow rate. Similar to the application of MPC to scheduling problems, the relaxation is used to avoid a high dimensional state space and thereby simplify the optimization problem [53, 78]. It is known that this adaptation changes the properties of the PN and that it is mostly suited for models with a large number of tokens [68, 75], which is why it is not suited for the problems discussed in this thesis.

10

In conclusion, there exists a solid basis for the application of MPC to various problem classes. A wide range of theoretical results can be exploited in the attempt to provide the desired guarantees for a given system class. With the help of simplifying relaxations, problems in production scheduling and for the control of PNs have been solved by means of MPC and theoretical guarantees can be given. For the control of JMPMSFs and for discrete PNs, the application of MPC to influence the discrete decisions in order to achieve a desired goal constitutes an open research question, which is considered in this work. As outlined in the following section, the goal of this thesis is to create a framework which allows to exploit the flexibility of an FMS motivated by the needs of Industry 4.0 and to provide guarantees through the application of MPC for its production scheduling.

## 1.3 Contribution and Outline of the Thesis

With this thesis, we contribute to the field of production scheduling for FMSs with methods from model predictive control (MPC) by developing a framework that links the description of the scheduling problem through automatic model generation with its solution methodology by means of MPC. In this framework, we unite the model generation and the solution of the scheduling problem based on the model. This is particularly useful to exploit the flexibility of the manufacturing system due to the possibility to react to changes in the real system by model adaptation or reactive scheduling. In the following, we outline the structure of this thesis and describe the main contributions of each chapter.

### Chapter 2 - Background and Preliminaries

In this chapter, we provide a brief explanation of the most relevant concepts from related fields for the contributions of this thesis. At first we introduce some basics from Industry 4.0, in particular the digital twin and skill-based engineering, which lead to the main assumptions taken throughout the rest of this work. As the second main field, we give an introduction to production scheduling. In this context, we explain the most important technical terms and lay a special emphasis on the systematic explanation of different kinds of flexibility in scheduling problems as well as the systematic classification of scheduling problems. In particular, we classify the problem class discussed in the remainder of this thesis and specify its flexibility. As the third main area covered in this work, we explain the most basic principles of Petri nets, which are the modeling formalism used to derive a mathematical representation of scheduling problems. The last research domain related to this thesis is MPC. Here we focus on the most basic theoretical concepts of MPC for linear discrete time systems, which prove to be sufficient for the developed results.

## Chapter 3 - Modeling of the Flexible Manufacturing System

In this chapter, we introduce the modeling and control framework that is developed in this thesis and formulate algorithms which allow to automatically generate mathematical models for the investigated class of scheduling problems. The objective of this chapter is to show how the creation of a mathematical model of an FMS and its model based control can be united in a common framework, in order to provide an integrated control engineering process.

The proposed framework builds upon on a simple formulation of the scheduling problem for an FMS and employs specifically developed algorithms to automate the generation of a mathematical model, which is the basis for the reactive scheduling by means of MPC. Consequently, it spans the whole range from system description until the model based control of the manufacturing system. As a starting point of the framework, we introduce a novel representation of a general scheduling problem. It is adjusted to the production of individual goods in Industry 4.0 by exploiting the modularity of a skill-based engineering setup through introducing a linking element between machines and products. Through its general nature it is able to represent most of the common production scheduling problems.

The automatic generation of a mathematical model for the provided problem description reduces the engineering effort for the application of the control framework. It exploits the modular structure of the description of the scheduling problem and leads to a linear discrete time model suited for the application of MPC. In order to formulate the model generation algorithms, we introduce an extension of the PN formalism, which allows to decouple autonomous evolutions of the system from conscious production decisions. We show that the PN model reflects the important characteristics of the scheduling problem. In particular, it provides the required flexibilities to represent Industry 4.0 scheduling problems based on the assumption of skill-based engineering. At the same time, the state space representation of the PN exhibits important properties for the scheduling of the manufacturing system by means of MPC.

In summary, the main contributions of this chapter are:

- The development of a framework that unites modeling and reactive scheduling.

- The formulation of the scheduling problem for an FMS in a modular way considering the principles of skill-based engineering.

- The proposition of automatic model generation algorithms that reduce the engineering effort.

- The introduction of an intuitive extension of the PN formalism to separate autonomous processes from conscious decisions.

- The investigation of the resulting mathematical model that captures the properties of the problem description and has a state space representation suited for MPC.

12

## Chapter 4 - Model Predictive Control for Production Scheduling

In this chapter, we develop two MPC schemes for the reactive scheduling of an FMS based on the model created in Chapter 3. The objective is to provide reactive scheduling schemes which are rigorously guaranteed to complete the scheduling problem in closed loop, incorporate the economic goal of the underlying manufacturing system and are able to react to changes in the system during runtime.

At first, we formulate an intuitive MPC scheme for the scheduling of the FMS based on the PN model. Due to the automatic generation of the model from the description of a production scheduling problem, it results in an economic MPC problem without predefined mode of operation. Its recursive feasibility is proven based on the properties that the model exhibits as a consequence of the automatic generation. For providing the guaranteed completion of the scheduling problem, we exploit the modular structure of the scheduling problem, which is conserved in its PN model. We first prove the completion of the smallest modules and infer the completion of the composite elements from the completion of all its parts until the entire production problem is guaranteed to be completed. Based on the insights from the analysis of the first MPC formulation, we formulate a slightly more advanced MPC scheme, which is computationally more efficient.

For both MPC schemes, we conjointly show how changes in the manufacturing system are handled through the feedback mechanism in the control loop or by exploiting the automated modeling framework developed in Chapter 3. We discuss the relationship between the cost function employed in the MPC problem, the elements of the PN and the elements of the original formulation of the scheduling problem. In particular, we highlight how the economic motivation of the scheduling problem can be respected while providing the completion guarantees. For the case of a linear cost function, we deduce simplifications in the computations required to provide the guarantees based on the previously given proofs. Finally, we show how existing techniques to guarantee the completion of the scheduling problem can be applied to the particular reactive scheduling formulation.

In summary, the main contributions of this chapter are:

- The presentation of two MPC schemes for the reactive scheduling of FMSs.

- Providing guarantees for the completion of the scheduling problem with the developed MPC schemes.

- The possibility of the MPC schemes to pursue the economic goal of the manufacturing system while guaranteeing its completion.

- The deduction of a simplified analysis for providing the guarantees in the case of a linear cost function.

## Chapter 5 - Numerical Examples

In this chapter, we employ the computationally more efficient MPC scheme proposed in Chapter 4 for the simulation of two exemplary scheduling problems from literature. The objective is to show how the MPC scheme can be used to solve scheduling problems that exhibit the characteristics of the problem class introduced in Chapter 3. We demonstrate that the proposed reactive scheduling scheme is computationally efficient, completes the scheduling problem and is able to react to changes in the system that occur during execution.

## Chapter 6 - Conclusion

In this chapter, we summarize the results provided in this thesis, discuss their advantages and limitations, and give an outlook on possible future work. We emphasize the requirements for the manufacturing system to meet the assumptions made in the description of the scheduling problem. We particularly stress how further extensions can be included in the proposed framework. Finally, we point out possible connections to further results from the literature on PNs and MPC that can be exploited to extend the results developed in this thesis.

## Appendix A - Proof of the Properties of the Automatically Generated Petri Net

The appendix contains the technical proof of the properties of the PN model resulting from the automatic model generation algorithms.

# Chapter 2

# Background and Preliminaries

The results in this thesis are based on fundamental concepts and results in production scheduling, modeling with Petri nets (PNs) and model predictive control (MPC). Therefore we briefly introduce them as preliminaries on which the rest of the thesis builds upon. Since many modeling and design choices of the developed framework are motivated by fundamental concepts in Industry 4.0, we start introducing them in Section 2.1, before presenting the fundamental preliminaries required for the rest of the thesis in the following sections. While the concepts of Industry 4.0 are explained to provide a comprehensive overview of the state of the art and ongoing developments, the further sections introduce well established terms and definitions. We start with a definition of the main terms from production scheduling that are relevant to this work and explain their dependencies in Section 2.2. In Section 2.3 we introduce Petri nets as the modeling tool that we use to represent the scheduling problem. We end in Section 2.4 with an introduction of the basics of MPC being the control concept employed for production scheduling. Parts of this section are based on and taken in parts literally from [88].

## 2.1 Concepts from Industry 4.0

The term "Industry 4.0" emerged as a paradigm in industrial production in 2013, when a dedicated working group in Germany presented a report in which they carved out the recent and anticipated changes in industry, economy and society [39]. Among other concepts, self-organized production systems were proposed, for which intelligent organization of assets and efficient communication between them are central aspects to achieve flexibility. To this end, the concept of *digital twin* was introduced and the *skill-based engineering* is being developed as a way to assign tasks to manufacturing units with ease. In this section, we will briefly introduce the core aspect of those concepts, which will be exploited in the formulation of the scheduling problem in Chapter 3.2 and the model creation in Chapter 3.3.

## Digital Twin

We take the concept of the digital twin as proposed by Grieves [27] as the basic principle to develop the MPC framework for production scheduling. It describes the notion of having a real-time synchronized virtual equivalent of the real world objects in a digital environment that can be used in various ways, for example for hardware-in-the-loop simulations or in decision-making processes [8, 27, 36, 61]. Assuming the existence of a digital twin means to assume that the required information of every physical system is always available in the corresponding digital twin and therefore can be used for the digital control of the manufacturing system. Production systems having such a digital representation and the capability to interact with their physical environment as well as with a digital control system and among one another by means of digital communication are called cyber-physical production systems [70]. The organization of the production, the manufacturing units, and the parts being produced is coordinated digitally by means of their digital twins, which comprises their current statuses, their abilities, and a digital documentation of their past. Starting from a model-based digital twin, it is enriched with data during its lifetime in order to increase its accuracy [36]. The up-to-date status information and the knowledge about the capabilities of the physical system represented in its digital twin allow for digital panning and control and is the prerequisite to formulate the online optimization problem introduced in Chapter 4. For a manufacturing unit, the digital twin particularly holds the information on manufacturing skills it can perform in the sense of skill-based engineering. This information needs to be provided to the control system, which can be done in the form of a digital directory as for example the "yellow pages for Industry 4.0" [44]. Such a digital directory serves as an organizational interface that allows to exploit the modularity and flexibility in the manufacturing system. Besides the digital twins of the manufacturing units, a digital twin is initialized for every customer specific order that is placed at the manufacturing system. From the digital twin of the products, especially the required steps for their production are of interest for our approach. As a result, the assignment between orders and manufacturing units is done based on their digital twins.

## Skill-Based Engineering

A promising approach to exploit the information provided in the digital twins is the concept of skill-based engineering, where capabilities of a manufacturing unit are abstracted to a higher level [17, 54]. On this abstraction level, the production system is represented in a modular fashion, in which the hardware and software functionalities are represented as distinct modules that offer specific skills and can be flexibly interconnected [23]. The products to be manufactured are also designed in a modular way and require certain skills for being produced. A matching between required skills for the production of a product and assured skills, which are provided by the machines, is performed. This concept can be seen as a variation and a further development of service-oriented automation, which is based on the concept of software services that allow flexible interconnection between software modules [62]. The formulation of the instructions to manufacture a product is focused on the workpiece that is being created, in the sense that it describes what to achieve with the workpiece but not the exact process required to achieve the goal [5]. Skill-based engineering uses the modularization of production processes in the form of skills as a means to facilitate process planning and automation in order to make it more flexible [69]. Due to its modularity, the single components in a skill-based architecture can be exchanged without losing functionality or requiring complex adaptions [17]. On the high abstraction level, the operator or the control software can access the skills of the machines and give the desired instructions, which are passed to the machines that interpret and execute the incoming commands according to their capabilities by means of a subordinate logic. This approach enables the flexible use of versatile manufacturing units in complex production scenarios. The goal of the scheduling scheme proposed in this thesis is to appropriately match the required skills and the assured skills on the higher abstraction level. On this level, the same task or skill might be implemented differently on different types of machines. For example, the tightening of a screw can be done with a tool that can rotate infinitely, or with a robotic arm that can only rotate by a limited angle and therefore needs to turn back and forth several times. Both implementations solve the same task and lead to the same result and therefore can be seen as the same skill, but both having different parameters in particular with respect to speed. In general, we consider the skill to execute a task in an abstract and universal way. For the problem formulation, which we will provide in Section 3.2, the skills are the central element for the assignment between jobs and manufacturing units. The proposed scheduling framework is an approach to systematically automate this assignment process.

## 2.2  Production Scheduling

As a brief introduction to the field of production scheduling, we first introduce the most important terms for the remainder of this thesis and explain their meaning. After that, in the scope of scheduling of flexible manufacturing systems (FMSs) in an Industry 4.0 context, we provide a more detailed overview on different types and notions of flexibility in the production scheduling context. In the last part of this section, we discuss a scheme from literature that is used to classify scheduling problems and classify the problem considered in the remainder of this thesis with respect to this scheme.

The process of *production scheduling* is the assignment of the operations, which have to be executed in a production system, to the available resources and a corresponding starting time with the goal to optimize one or more objectives [6, 31, 66]. The result of production scheduling is a *schedule*, which is an assignment of operations to machines and a starting time [6]. These definitions rely on the terms production system, operation and resource, which have to be defined in this particular context.

A *production system*, which we also call *manufacturing system* interchangeably, is considered to be an abstract entity that encompasses various resources and has the goal to create goods and/or services [33]. In general, resources are considered to be the "five M's" being men, machines, methods, materials and money [33].

In the context of production scheduling considered in this thesis, we particularly focus on the resource *machines*, which we refer to synonymously as *manufacturing units*. In the abstract representation of the manufacturing system, also a workstation requiring a human worker can be abstracted to be a "machine". Hence, in the production scheduling considered in this thesis, operations are assigned to machines.

As a flexible manufacturing system, we consider a special type of production system that provides flexibility on the hardware level and with respect to the interconnections of its resources. It is composed of elements, e.g., machines, transportation units or work stations, that either do not have a dedicated purpose and therefore can be used for different use cases, or they are able to process a variety of part types through variation in their interconnection, or they offer both possibilities [90]. It is the foundation of an intelligent manufacturing system, in which the flexibility of the FMS is provided to a control system that is able take advantage of it through suitable control actions.

An *operation* is a processing step that requires a certain amount of time on a machine for being completed [65]. Through the division into several operations, the work to be done for the creation of a product is clustered into distinct steps. Each operation or processing step has unifying properties internally and distinguishing properties with respect to other operations.

Operations of the same kind that are required for the production of different products are considered to be the same *task* in this thesis. If the same task is required for the production of different goods, the specific task for the creation of a particular good is an operation. In the context of skill-based engineering, the term 'skill' is used for what is called 'task' in the scheduling problem. The distinction between required skills to produce a product and assured skills of a machine is handled implicitly by assigning a task to the instruction for the production of the product, or to the capabilities of a machine that can execute the task, respectivly. If the production of a product requires a particular skill, the task representing this skill is assigned to the instructions to produce the product in the description of the production instructions in the scheduling problem. In the same way, a task is assigned to the capabilities of a machine if the machine provides the skill to execute the task.

The set of instructions to produce a product is called *job* in the context of production scheduling. A job consists of multiple tasks that have to be completed in order to complete the job. A task as a processing step in the production of some specific product, which is described as a particular job, is referred to as an operation. Two operations of two different jobs are considered to be the same task if they have the same instructions, requirements, etc. In these terms, the production scheduling is the assignment of *tasks* that are part of *jobs* to *machines*. Those three elements are the central elements in the problem formulation in Chapter 3.2.

## Flexibility in Production Scheduling

Since flexibility plays a crucial role in the development and implementation of Industry 4.0 [39, 84, 90], we introduce five relevant types of flexibility in production scheduling. This list is not exhaustive but focuses on the flexibilities considered in the remainder of this thesis. More in-depth analyses of different types of flexibility in manufacturing are, among others, provided in [4, 37, 74].

In general, we consider flexibility of a manufacturing system as its ability to adapt, or to be adapted, as a response to changing external conditions [4]. An operator or operating system needs to have the possibility to exploit the flexibility in order to improve the system behavior with respect to some desired criteria, for example, the efficiency and profitability of the manufacturing system. The most significant types of flexibility of the considered description of an FMS are

(F1) the possibility to execute the same task on different machines,

(F2) the ability of one machine to execute different tasks,

(F3) the possibility to change the sequence in which the tasks in one job are executed,

(F4) the possibility to change the sequence in which different operations are executed on a machine,

(F5) the possibility to execute different types of jobs in the same manufacturing system.

In the search of a unified taxonomy, Sethi and Sethi [74] identified more than fifty different terms for various types of flexibility and there are different terms for the same type of flexibility as well as different meanings of the same terms. This is why we do not set specific terms for the flexibilities (F1)–(F5), but rather refer to their numbers (F1)–(F5) when referring to the different types of flexibility. With respect to the taxonomy of Sethi and Sethi [74], who distinguish eleven types of flexibility, the description of an FMS in Section 3.2 offers routing flexibility (F1), machine flexibility (F2), operation flexibility (F3) and (F4), and process, product and production flexibility (F5). Implicitly, material handling flexibility is also available, since the jobs are assumed to be transferable from one machine to another. The setup in Section 3.2 also allows for expansion flexibility, which describes the ability to introduce new machines to the system, due to the automatic generation of the scheduling problem in Section 3.3.2.

## Classification of Scheduling Problems

A common classification of scheduling problems was introduced by Graham et al. [26] and further refined since then [10, 35, 65]. It systematically categorizes various scheduling problems by their machine environment, their job characteristics and their optimality criterion. For the classification of the problem class discussed in this thesis, we mainly focus on the machine environment since the job characteristics in the sense of the classification scheme are not completely specified and the optimality criterion depends on the application case and only needs to have the properties required for the proofs in Section 4.

The machine environment of the scheduling problem that we will describe in Section 3.2 is a generalization of the job shop (JS), which is one of the most common scheduling problems in literature [10, 65]. It is defined by a given number of machines and a given number of jobs. Each job has a well defined production plan consisting of a list of tasks that have to be executed in a predetermined sequence. Every task has to be executed by a dedicated machine, which is why the sequence of tasks is also described as a route that the job has to follow through the production system. There is a strict linear precedence relation between the tasks in a job. The jobs can be different from each other and therefore generally have different sets of tasks. With respect to the types of flexibility defined in Section 2.2, the JS offers the flexibilities (F4) and (F5). In the most common descriptions of a JS, it is not specified whether the Flexibility (F2) is available and the machines can execute different tasks.

From the JS described above, the partial job shop (PJS) and the job shop with multi-murpose machines (JMPM) are deduced. They generalize the JS in different ways.

In the PJS, which was introduced by Nasiri and Kianfar [60] and is rarely considered in literature [94], sequence flexibility (F3) is introduced with respect to the JS. The fixed sequence of tasks in a job that is present in the JS is relaxed such that arbitrary precedence relations between tasks of the same job in the form of an acyclic graph are allowed. The machine flexibility (F1) to choose a machine on which a certain operation will be executed is not available in the PJS.

In the JMPM, machine flexibility (F1) is introduced with respect to the JS by allowing every task to be executed by one of multiple possible machines [10]. The tasks in a job, however, still have a predetermined sequential order in which they have to be executed [9, 13]. In literature, the JMPM is the most common definition of what is frequently called flexible job shop (FJS) [13, 24, 80]. Although the FJS is frequently discussed in literature, it is not always defined in the same way. In a more restrictive definition by Pinedo [65], the FJS consists of distinct work centers with identical parallel machines.

The machine environment in the scheduling problem that we will describe in Section 3.2 unites the flexibilities of the PJS and the JMPM with respect to the ordinary JS and therefore is a generalization of both of them. It is a JS with additional machine flexibility (F1) and sequence flexibility (F3). Therefore, in order to arrive at an understandable denomination that speaks for itself, we follow the definition of the JMPM and categorize the machine environment discussed in the remainder of this thesis as *job shop with multi-purpose machines and sequence flexibility (JMPMSF)*. In the literature no unique name was given to the JMPMSF. Among other names, it is called *extended FJS* [6], or *FJS with process plan flexibility* [63]. In contrast to the definition of the JMPM by Brucker [10], we not only explicitly introduce the Flexibility (F1), meaning that a single task can be executed on different machines, but also that the reverse. We explicitly include the possibility that a single machine can execute different tasks (F2). The systematic classification of the considered class of scheduling problems and the explicit definition of the available flexibilities helps with the development of automatic model generation algorithms in Section 3.3.

## 2.3 Petri Nets

In order to have a foundation for describing the automatic generation of a model of the JMPMSF, we formally introduce Petri nets as a modeling tool. In this context, we only introduce the type of PN as introduced by Petri [64], which is the basis for many extensions and variations that build upon it, but are beyond the scope of this thesis. Since we mainly use the algebraic representation of PNs in the remainder of this work, we omit many common definitions related to PNs and only introduce the most relevant terms and concepts in this section. In this introduction we follow some basic descriptions from [11, 25, 73], where further modeling and analysis techniques based on PNs can be found.

**Definition 1** (Petri net). *A Petri net is a weighted bipartite directed marked graph described by a tuple $PN = (\mathbb{P}, \mathbb{T}, \mathbb{E}, w, x^0)$, where*

- $\mathbb{P} = \{P_1, ..., P_n\}$ *is a finite set of* places, $n \in \mathbb{N}_{>0}$, *graphically represented as circles,*

- $\mathbb{T} = \{T_1, ..., T_m\}$ *is a finite set of* transitions, $m \in \mathbb{N}_{>0}$, *graphically represented as bars,*

- $\mathbb{E} \subseteq (\mathbb{P} \times \mathbb{T}) \bigcup (\mathbb{T} \times \mathbb{P})$ *is a set of* arcs *from places to transitions and from transitions to places, graphically represented as arrows,*

- $w : \mathbb{E} \to \mathbb{N}_{>0}$ *is an* arc weight function, *graphically represented as numbers labeling the arcs (if an arc has the weight 1 it is not labeled), and*

- $x^0 \in \mathbb{N}^n$ *is the* initial marking *of the Petri net, from now on called* initial state; *the initial state of a place $P_i$ is graphically indicated by $x_i^0$ dots ("tokens") in the circle corresponding to $P_i$.*

The dynamics of a PN is driven by the *firing* of the transitions and captured by the evolution of the state vector $x(k) \in \mathbb{X} \subseteq \mathbb{N}^n$ at the time instants $k \in \mathbb{N}$ and is initialized with $x(0) = x^0$. The firing of a transition $T$ removes tokens according to the weights $w(P_i, T)$ of the arcs $(P_i, T) \in \mathbb{E}$ from its input places $P_i \in \mathbb{P}$ and adds tokens to its output places $P_o \in \mathbb{P}$ according to the weights $w(T, P_o)$ of the arcs $(T, P_o) \in \mathbb{E}$. The number of times each transition fires at instant $k$ is expressed with the *firing count vector* $u(k) \in \mathbb{U} \subseteq \mathbb{N}^m$. We introduce the incidence matrix $B = B^+ - B^-$, in

21

which every column corresponds to one transition $T_j$ and which is composed of the two matrices $B^+ \in \mathbb{N}^{n \times m}$ and $B^- \in \mathbb{N}^{n \times m}$ that are defined as

$$B_{i,j}^+ := \begin{cases} w(T_j, P_i) & \text{if} \quad (T_j, P_i) \in \mathbb{E} \\ 0 & \text{else} \end{cases} \quad \text{and } B_{i,j}^- := \begin{cases} w(P_i, T_j) & \text{if} \quad (P_i, T_j) \in \mathbb{E} \\ 0 & \text{else} \end{cases}. \quad (2.1)$$

We can now compactly describe the PN dynamics by

$$x(k+1) = x(k) + Bu(k), \qquad x(0) = x^0. \quad (2.2)$$

As the places are not allowed to have a negative number of tokens, the Petri net dynamics needs to be further restricted by only allowing transitions to fire if they do not generate negative tokens in any place. We say a transition $T$ is *enabled* at a state $x(k)$ if all of its input places $P_i$ connected to $T$ through an arc $(P_i, T) \in \mathbb{E}$ hold at least as many tokens as the weight $w(P_i, T)$ of this arc, i.e., $x_i \geq w(P_i, T)$ [25]. Two transitions, $T_i$ and $T_j$, might be concurrently enabled at state $x(k)$, but if they both fire simultaneously they would consume the same token, resulting in a negative entry in the state vector $x(k+1)$. This situation is called *conflict* between the transitions $T_i$ and $T_j$ [73, Chap. 10]. To prevent negative tokens in this case, a non-negativity condition on the basis of the firing count vector $u(k)$ is introduced. With the matrix $B^-$ and the state vector $x(k)$ all allowed firing count vectors $u(k)$ must satisfy the non-negativity constraint

$$0 \leq x(k) - B^- u(k). \quad (2.3)$$

The non-negativity constraint (2.3) only considers the input-arcs to the transitions, which are captured in the matrix $B^-$. This is important in order to only allow circular firings if all transitions in the circle are concurrently enabled. If instead of $B^-$ the incidence matrix $B$ was used, a circular firing would also be possible, even if no single token was present in the circle.

This rudimentary introduction of the very basic form of a PN and its algebraic description in the form of the state dynamics (2.2) and the non-negativity constraint (2.3) is sufficient to automatically generate a model for the JMPMSF description of an FMS in Section 3.3. In this course we slightly extend the algebraic description of a Petri net (2.2) in order to increase its descriptive power. The resulting PN model will be used in Chapter 4 to derive an MPC based reactive scheduling scheme for FMSs.

## 2.4 Model Predictive Control

As an introduction to MPC for the remainder of this thesis, the most universal aspects are sufficient, which are introduced in [28, Chap. 1] and [67, Chap. 2] in a more general context.

For the MPC problems considered in this thesis, we assume that the systems to be controlled are modeled by a linear time invariant discrete time dynamical system of the form

$$x(k+1) = Ax(k) + Bu(k), \quad x(0) = x^0, \tag{2.4}$$

where $k \in \mathbb{N}$ describes the considered time instant, $x \in \mathbb{X} \subseteq \mathbb{R}^n$ is the state, which is restricted to the state space $\mathbb{X}$, and $u \in \mathbb{U} \subseteq \mathbb{R}^m$ is the input, which is restricted to the input space $\mathbb{U}$. The evolution of the system is described through the matrix $A \in \mathbb{R}^{n \times n}$ and the input matrix $B \in \mathbb{R}^{n \times m}$ starting from a given initial state $x^0 \in \mathbb{X}$. The system needs to respect the combined state and input constraints

$$(x(k), u(k)) \in \mathbb{Z} \subseteq \mathbb{X} \times \mathbb{U} \tag{2.5}$$

at every time instant $k \in \mathbb{N}$, where $\mathbb{Z}$ is a given constraint set.

In the general MPC setup, the system is controlled by solving a finite horizon optimal control problem at each time instant $k$ and only applying the optimal first input to the system. In the optimal control problem

$$\underset{u(\cdot|k)}{\text{minimize}} \quad \sum_{\bar{k}=0}^{N-1} c\left(x\left(\bar{k}|k\right), u\left(\bar{k}|k\right)\right) + V_{\mathrm{f}}\left(x(N|k)\right) \tag{2.6a}$$

$$\text{subject to} \quad x\left(\bar{k}+1|k\right) = Ax\left(\bar{k}|k\right) + Bu\left(\bar{k}|k\right), \tag{2.6b}$$

$$\left(x\left(\bar{k}|k\right), u\left(\bar{k}|k\right)\right) \in \mathbb{Z}, \quad \text{for } \bar{k} = 0, \ldots, N-1, \tag{2.6c}$$

$$x\left(0|k\right) = x\left(k\right), \tag{2.6d}$$

a cost function $c : \mathbb{X} \times \mathbb{U} \to \mathbb{R}$, which is commonly called *stage cost* in MPC literature, is minimized. The optimization variables are the elements of the trajectory of planned inputs $u\left(\cdot|k\right) = \left(x(0|k), \ldots, u(N-1|k)\right)$. During the optimization, initialized with the measured state $x(0|k) = x(k)$, the system dynamics (2.6b) and the state and input constraints (2.6c) are respected. The system behavior is predicted $N$ steps into the future, where we call $N \in \mathbb{N}_{>0}$ the prediction horizon, leading to the predicted state trajectory $x\left(\cdot|k\right) = \left(x(0|k), \ldots, x(N|k)\right)$. In this notation, $u\left(\bar{k}|k\right)$ denotes the input that is planned to be applied at the time instant $k + \bar{k}$ in the prediction based on the measured state $x(k)$. Analogously, $x\left(\bar{k}|k\right)$ describes the state predicted for the time instant $k + \bar{k}$ based on the measured state $x(k)$ by applying the predicted inputs $u(\tilde{k}|k)$, $\tilde{k} = \{0, 1, \ldots, \bar{k}\}$ to the system dynamics (2.4). The final state $x(N|k)$ in the predicted sequence is weighted with a terminal cost function $V_{\mathrm{f}} : \mathbb{R}^n \to \mathbb{R}$. This is not required in all MPC formulations, but a common way to ensure some desired properties of the resulting closed loop system [67].

We assume that the optimization (2.6) is carried out until an optimal input trajectory $u^*(\cdot|k)$ is found. In contrast to many approaches from MPC literature, where uniqueness of the optimizer is achieved by particular design choices, it is not assumed or required for our purposes and an arbitrary input trajectory minimizing (2.6) is chosen as $u^*(\cdot|k)$. The control loop is closed by applying the first input of the optimal planned input sequence to the system, i.e., $u(k) = u^*(0|k)$. In order to introduce a feedback mechanism, at the next time instant $k+1$ the state of the system $x(k+1)$ is measured and the optimization (2.6) is repeated with the initial condition $x(0|k+1) = x(k+1)$. This can be expressed in the basic MPC Algorithm 1.

---

**Algorithm 1:** Basic linear MPC algorithm.

**parameters:** The model (2.4) of a dynamical system to control, a stage cost function $c : \mathbb{R}^n \times \mathbb{R}^m \to \mathbb{R}$, a terminal cost function $V_f : \mathbb{R}^n \to \mathbb{R}$, a constraint set $\mathbb{Z} \subseteq \mathbb{X}^n \times \mathbb{U}$ and a prediction horizon $N \in \mathbb{N}_{>0}$

1 **for** $k = 0, 1, \dots$ **do**
2     Measure the true state $x(k)$ of the real system;
3     Solve (2.6) with $x(0|k) = x(k)$;
4     Apply $u(k) = u^*(0|k)$ to the real system;
5 **end**

---

The parameters in Algorithm 1 are either given or have to be determined offline beforehand accoring to the desired control goals. The resulting closed loop form applying Algorithm 1 can be expressed as

$$x(k+1) = Ax + Bu^*(0|k) = x^*(1|k). \tag{2.7}$$

Due to its nature as a feedback control method, the main applications of MPC are in the stabilization of a desired set point $(x^s, u^s) \in \mathbb{Z}$ or tracking some reference trajectory $(x^r(k), u^r(k)) \in \mathbb{Z}$, both while satisfying the constraints. Beyond that, in the field of *economic* MPC, higher level goals expressed through the cost function are addressed [67, Chap. 2.8]. While in stabilizing MPC and in tracking MPC, for the most part, the cost function is the means to achieve stability and convergence of the system to the desired state, in economic MPC the minimization of the cost function is the desired control goal. In this case the closed loop system behavior is not defined beforehand. To provide guarantees for the behavior of the closed loop system (2.7) controlled by the MPC (2.6), assumptions on the cost function $c(x, u)$, the system dynamics (2.4) and the constraints $\mathbb{Z}$ are made, or those MPC ingredients have to be suitably chosen. Similarly, they need to satisfy certain properties that guarantee the feasibility of the MPC problem (2.6), such that the model predictive controller is a well defined control law. In Chapter 4, we develop specific criteria of the MPC ingredients, which have to be satisfied to prove the desired closed loop properties of the considered production scheduling scheme. Our approaches are related to schemes in which asymptotic stability of the origin is guaranteed by a sufficiently long prediction horizon, cf. [28, Chap. 6] and the references therein, and to schemes in which it is achieved through a suitably chosen terminal cost [28, Chap. 5] . Limon et al. [46] show the relation between terminal cost $V_f(x(N|k))$, length of the prediction horizon $N$ and the performance of the closed loop system with respect to the stage cost $c(x, u)$.

This consideration plays an important role in the evaluation of the MPC schemes presented in Chapter 4. Since we built upon basic MPC principles, we remain in this introduction on this abstract level and refer to the standard literature, e.g., [28, 67], for further details on existing MPC techniques.

## 2.5 Summary

In this chapter, we introduced the most relevant concepts from the four main areas that are related to this thesis. By starting with the foundation of Industry 4.0 and in particular the digital twin and skill-based engineering, we set the basis for the following investigations. For the development of a novel scheduling framework, we explained the most relevant terms from the field of production scheduling and defined important flexibility notions, which help to classify the problem class covered in this thesis. Based on that, we introduced the job shop with multi-purpose machines and sequence flexibility (JMPMSF) as a general class of scheduling problems. As a tool to create models for scheduling problems from the class of JMPMSF, we introduced the most fundamental terms of Petri nets. In this regard, we focused on their very basic definition. Finally, we introduced MPC as the control scheme that will be used for the scheduling of the JMPMSF based on its PN model. We explained the basic MPC algorithm as feedback law and the main elements involved in its formulation. The reactive scheduling framework presented in the following chapters relies on the concepts of Industry 4.0 as preliminaries. It considers the general class of JMPMSF scheduling problems, from which a PN model is built as the basis for an MPC scheme. This MPC scheme can be employed to solve the scheduling problem in light of Industry 4.0 scenarios.

# Chapter 3

# Modeling of the Flexible Manufacturing System

In this chapter, we describe an automatic model generation scheme for a flexible manufacturing system (FMS). In Section 3.1, an overview on the framework is provided, the steps towards a model for the reactive scheduling mechanism are sketched and we briefly introduce how it will be used. The job shop with multi-purpose machines and sequence flexibility (JMPMSF) as a structured description of the FMS is provided in Section 3.2. At first, we introduce the basic elements of the JMPMSF and explain how they relate to one another. In a second step, we describe which flexibilities the JMPMSF provides and which restrictions there are between its elements. We end Section 3.2 with a description of possible scheduling objectives that are suited for the reactive scheduling scheme. From the JMPMSF, a Petri net (PN) model is automatically generated with algorithms presented in Section 3.3. For the model generation, we slightly modify the general notion of PNs introduced in Section 2.3, such that its state space description has the form of a linear time invariant discrete time system, which is suited for model predictive control (MPC) as introduced in Section 2.4. At the end of the chapter, we describe and analyze the properties of the resulting PN model and its state space description and relate it to similar scheduling problems.

## 3.1 Framework Overview

The framework presented in this thesis is specifically designed for the model based production scheduling in FMSs. It does not only automate the control of an FMS but also the modeling process as depicted in Figure 3.1. Assuming a given real world FMS, we explain how to formulate the scheduling problem as basis for the model based reactive scheduling scheme. Based on this problem description, the model generation for the model predictive production scheduling is proposed by means of an automated process. In this section, we give a brief overview on the complete framework and highlight the key features of the intermediate steps.

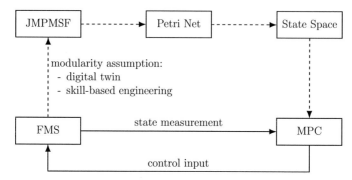

Figure 3.1: Illustration of the modeling and control framework for the model predictive scheduling of flexible manufacturing systems. The modeling process is sketched with dashed lines and the resulting control loop is depicted with solid lines.

## Job Shop with Multi-Purpose Machines and Sequence Flexibility

The job shop with multi-purpose machines and sequence flexibility (JMPMSF) is the starting point for the automatic model generation process. A main requirement to start the automatic model generation for the FMS is that the underlying manufacturing system is built in a modular fashion and the production of the goods can be divided modularly into distinct production steps. This represents the notion of skill-based engineering as introduced in Section 2.1. In the JMPMSF formulation, the planned production of a product is formulated as a job that consists of required tasks. The tasks to manufacture the product require skills of the machines such that they are able to execute the required production steps. The tasks have to be executed to complete the job and finalize the corresponding product. The different tasks in one job for the production of some good depend on one another, meaning that some tasks have to be completed before others can be started. This restricts the possible evolution of the production process. We assume that the tasks in each job and their dependencies are known. Similarly, the machines in the manufacturing system offer the skills to execute the tasks and we assume the skills of each machine are known. With the tasks as linking element, the jobs and machines are matched. We formulate this matching as a scheduling problem between the *required skills* to produce a product and the *assured skills* provided by the machines.

Since the production planning is performed in a virtual environment, the up-to-date status information about the products and machines needs to be available. Therefore, we assume that there exist digital twins as described in Section 2.1 for the production units as well as for the products being produced. For the products, this means that the required production steps are available at the initialization of their digital twins.

During production, the production steps that are already completed are continuously updated in the digital twins of the products. This updating process is part of the state measurements required in the feedback control loop as depicted in Figure 3.1. For the production units, the availability of a digital twin means that their capabilities are known in the virtual environment. During runtime, the machines are monitored and it is known whether they are idle and available, working or defective. Their status information is assumed to be available as feedback in the control loop. Only then, the reactive scheduling scheme can make production decisions based on the most up-to-date information of the production system. Further details on the JMPMSF representation of the FMS are provided in Section 3.2.

**Petri Net**

For the automatic generation of the model from the JMPMSF problem description, the modularity of the skill-based FMS representation is exploited. The different elements of the JMPMSF are systematically analyzed and transformed into a Petri net (PN) representation. Due to the automatic model generation, only the problem description in the form of a JMPMSF needs to be done manually, while the modeling process is executed without requiring human intervention. In this process, the elements of the JMPMSF are combined according to their dependencies and PN elements are created. In the PN representation, the products and production units are linked based on their required and assured skills, i.e., the tasks assigned to them. In the PN, every matching combination of machine, task and job is introduced as a sequence of places and transitions that represents one possible execution of the task, which is necessary for a job, on some specific machine. A matching combination is characterized by a task that is on the one hand required in the job and on the other hand the machine is capable to execute it. In this way, all possible production sequences are represented in the PN. Since only one production sequence needs to be realized at the end, the choice between different production sequences needs to be implemented in the PN. Once the decision is made and some specific production process is started, the production process itself needs to run autonomously and independently of any further user decisions. To distinguish conscious decisions from autonomous behavior, we introduce a novel distinction between two kinds of transitions, the *controlled* transitions and the *independent* ones. An independent transition is related to a single input place in a way that allows its firing independently of any other transition. Through this restriction, the PN is guaranteed to have a well defined behavior if the independent transitions fire as soon as they are enabled. The controlled transitions can have an arbitrary relation to their input places, requiring to check whether they are allowed to fire. This distinction increases the expressive power of the PN and its algebraic description and makes it easier to use the PN for MPC.

**Feedback Control Loop**

In the feedback control loop, an MPC scheme based on the PN model is used for the real-time scheduling of the FMS. To this end, on the one hand the inputs in the form of scheduling decisions need to be computed and assigned to the manufacturing system, and on the other hand the status of the FMS needs to be measured and fed back to the controller. We assume that this is possible on the basis of the digital twins as real-time synchronized copies of the real objects. Through the digital twins, the up-to-date status is available in the digital control system. With this information, which is assumed to be correct, an MPC optimization problem is solved and its solution is the control input that is assigned to the manufacturing system. This closed loop system has the ability to react to changes and disturbances by perceiving them through the measurements and exploiting the flexibility of the FMS in the MPC to initiate an appropriate reaction. In Chapter 4, we will describe this in more detail. We will prove that the MPC problem is always feasible and that the scheduling problem is guaranteed to be completed by the MPC in closed loop if a simple and intuitive condition is fulfilled.

## 3.2 The Job Shop with Multi-Purpose Machines and Sequence Flexibility

In this section, we introduce the job shop with multi-purpose machines and sequence flexibility (JMPMSF) in the formulation that was originally described in [87]. It is a general case of a flexible job shop (FJS) and describes a flexible manufacturing system with a given set of manufacturing units in which customer specific orders are fulfilled. Since it does not have many restrictions and constitutes a very flexible problem in the field of production scheduling, it serves well as a general problem class to develop a versatile reactive scheduling technique in Chapter 4. In Section 3.2.1, the main components and their interconnections are introduced. In Section 3.2.2, we further specify the JMPMSF by describing the restrictions in the scheduling problem that arise from real world restrictions of the production system and analyze the flexibility that the resulting problem description provides. Finally Section 3.2.3 describes the class of objectives that the proposed scheduling framework is applicable for. The results presented in this section are based on [87] and [88] and taken in parts literally from [88].

### 3.2.1 Basic Elements of the JMPMSF

For the modeling of an FMS in which customer specific orders are fulfilled, we assume to have the necessary information on the manufacturing system readily available. It is assumed that each order of a customer specific product constitutes a well defined and possibly unique job that needs to be completed by the manufacturing system.

Consequently, the required information are the basic elements of the manufacturing problem, which are

- a set $\mathcal{T} = \{\tau_1, \ldots, \tau_{n_T}\}$ of tasks that can be executed in the manufacturing system,

- a set $\mathcal{M} = \{M_1, \ldots, M_{n_M}\}$ of manufacturing units (robots, machines, automated guided vehicles, etc.) which are called machines for brevity in the rest of this thesis; every machine $M$ can only execute a subset of tasks, which is specified as $\mathcal{T}_M \subseteq \mathcal{T}$,

- a set $\mathcal{J} = \{J_1, \ldots, J_{n_J}\}$ of jobs that need to be fulfilled by the manufacturing system; every job $J$ consists of a set $\mathcal{T}_J \subseteq \mathcal{T}$ of tasks.

The purpose of the manufacturing system is to complete the jobs $J \in \mathcal{J}$ with the available machine pool $\mathcal{M}$. The assignment between jobs and machines is done based on the tasks $\tau \in \mathcal{T}$, which are the linking element between them. In the sense of skill-based engineering [54], the manufacturing units have the required skills to execute tasks, which are the basic building blocks for the jobs that have to be completed by the manufacturing system. A job is completed once all its tasks have been completed.

As different jobs $J$ and $J'$, which represent the instructions for the production of different goods, may share the same production steps, their sets $\mathcal{T}_J$ and $\mathcal{T}_{J'}$ may contain the same elements. If we refer to a task $\tau \in \mathcal{T}_J$ of a specific job $J$, the pair $(\tau, J)$ of task and job is called *operation* $O = (\tau, J)$. With this terminology, the set of operations $\mathcal{O}$ is defined as $\mathcal{O} = \{(\tau, J) | \tau \in \mathcal{T}_J\} \subseteq (\mathcal{T} \times \mathcal{J})$. The term operation is mainly used to abbreviate certain statements, but in many cases we rather refer to the pair of task and job for the sake of clarity.

In the same way different jobs might share the same tasks, also different machines $M$ and $M'$ might be able to execute the same tasks and hence the corresponding sets $\mathcal{T}_M$ and $\mathcal{T}_{M'}$ may have elements in common. This implements the Flexibility (F1) and might even be the case if the machines are of different types. For example, a machine $M$ with five axes and a machine $M'$ with three axes might both be able to execute a task $\tau$ that only requires three degrees of freedom [23]. In this case, the task $\tau$ is in the set of executable tasks of both machines, i.e., $\tau \in \mathcal{T}_M$ and $\tau \in \mathcal{T}_{M'}$. In the framework of skill-based engineering, the fact that two machines are able to execute the same task means that they both offer the required skill for the execution of the task. This, however, does not imply that both machines execute this task in the same way [17, 79]. The machines might be of different kinds and it is not even required that the machines have some specific tool, as the introduction of a multi-purpose machine by Brucker [10] suggests. As the way in which a task is executed may vary from machine to machine, the production time $t_P(M, \tau) \in \mathbb{R}_{>0}$ describing the amount of time that machine $M \in \mathcal{M}$ needs to execute task $\tau \in \mathcal{T}_M$ not only depends on the task but also on the machine. The production times $t_P(M, \tau)$ are assumed to be known and fixed for every valid combination of machine and task, i.e., for all $\tau \in \mathcal{T}_M$ with $M \in \mathcal{M}$.

In scheduling literature, the assignment between tasks and the machines that are able to process them is viewed from the perspective of the task [10]. Each task, has a corresponding set of machines that are able to process it. The machines themselves are not inherently described with a set of tasks that they can execute. In an Industry 4.0 scenario that is considered over a possibly infinite time horizon, this notion does not seem suited. As described in the context of skill-based engineering [54], the skills that each machine can perform need to be known and therefore either the tasks that each machine can execute are already known, or they can be deduced based on the machine's capabilities. When jobs are allowed to enter the manufacturing system and are assumed to be completed and removed at some later point in time, the implications of new jobs entering the manufacturing system and completed ones being removed from it need to be considered. Even if a new job requires a novel task $\tau^*$ that has never been executed in the manufacturing system before, it is reasonable to check which machines $M$ are able to execute it and add the novel task $\tau^*$ to their sets of executable tasks $\mathcal{T}_M = \mathcal{T}_M^{(\text{old})} \cup \tau^*$. This new task can then either be remembered for the case that a future job also requires the same task in its production process, or it is discarded after the job is finished. The part of the production problem that can be considered to be persistent over time are the machines in the manufacturing system. Thus, an assignment between machines and their production capabilities as introduced above, i.e., the skills to execute tasks, will not be subject to many changes. In contrast to that, in the common method from literature the jobs, or rather their tasks, have a corresponding set of machines that can execute them. Those sets need to be reintroduced with every new job. The process of determining which machines are able to execute some specific task needs to be executed for all tasks of every new job, which requires additional effort.

The basic elements tasks, machines and jobs provide an abstract description of the manufacturing problem which can be refined to describe several more specific scenarios. In the next section, we provide the means to formulate restrictions on the production problem based on the elements introduced above, which is necessary to describe realistic production scenarios. Depending on the amount of restrictions present in the particular production scenario, it offers a variety of different flexibilities. We want to exploit those flexibilities and combine the skills of the available manufacturing units in order to implicitly create and describe various possible production plans. The availability of multiple production plans is the basis for finding an optimal plan with respect to a given objective in Chapter 4. In the next section, we provide a structured description of the restrictions and flexibilities.

## 3.2.2 Restrictions and Flexibility in the Scheduling Problem

In the considered scheduling problem, there are restrictions concerning the three basic elements, i.e., tasks, machines, and jobs. To begin with, the sequence of tasks in a job is generally free. However, as there might be dependencies between the different steps in the production process of a product, these dependencies need to be considered in the manufacturing problem such that only production sequences are scheduled which can be executed in the real production process. This is achieved through the following four restrictions in the scheduling problem related to the tasks:

(R1) A task $\tau$ might not be independent, but require all tasks $\tau' \in \mathcal{T}_\tau$ in the set of its required tasks $\mathcal{T}_\tau \subset \mathcal{T}$ to be finished before it can be started.

(R2) A task $\tau \in \mathcal{T}_J$ in a job $J$ can only depend on tasks $\tau' \in \mathcal{T}_J$ in the same job $J$, i.e., $\mathcal{T}_\tau \subset \mathcal{T}_J$ for every task $\tau \in \mathcal{T}_J$. If a task $\tau$ depends on another task $\tau'$, meaning that $\tau' \in \mathcal{T}_\tau$, $\tau \in \mathcal{T}_J$, then the operations $O = (\tau, J)$ and $O' = (\tau', J)$ exist and it is said that $O$ depends on $O'$. Operations $O = (\tau, J)$ and $\bar{O} = (\tau', \bar{J})$ in different jobs $J$ and $\bar{J}$ are independent from one another.

(R3) Two tasks $\tau$ and $\tau'$ must not be mutually dependent.

(R4) Preemption of tasks is not allowed, i.e., once a task has been started it must be completed without interruption.

The restriction to a non-preemptive setup is the mostly considered case in literature and therefore adopted in the considered problem [13, 15, 24, 80, 91]. In the description of the dependencies of a task $\tau$ on other tasks (R1), only *direct* dependencies need to be provided in the sets $\mathcal{T}_\tau$. Indirect dependencies with one or more other tasks in between can be determined with the recursive function $\Phi$ described in Algorithm 2. The resulting set of all tasks on which a task $\tau$ depends directly or indirectly is denoted with $\bar{\mathcal{T}}_\tau$.

---
**Algorithm 2:** Determine indirect dependencies.
---
1 **Function** $\bar{\mathcal{T}}_\tau = \Phi(\mathcal{T}_\tau)$
2   | $\bar{\mathcal{T}}_\tau = \mathcal{T}_\tau$ ;
3   | **for** *every Task* $\bar{\tau} \in \mathcal{T}_\tau$ **do**
4   |   | $\bar{\mathcal{T}}_\tau = \bar{\mathcal{T}}_\tau \cup \Phi(\mathcal{T}_{\bar{\tau}})$ ;
5   | **end**
6   | **return** $\bar{\mathcal{T}}_\tau$
7 **end**

---

From the different sets $\mathcal{T}_\tau$ of the tasks $\tau \in \mathcal{T}_J$ in a job $J$, a precedence graph $G_J = (\mathcal{T}_J, E_J)$ can be generated, as we illustrate in Example 1. In this graph, the precedence relations between the tasks in a job are represented by considering the tasks $\tau \in \mathcal{T}_J$ as nodes and constructing the directed edges $E_J = \{(\tau', \tau) | \tau' \in \mathcal{T}_\tau, \tau \in \mathcal{T}_J\}$ according to the precedence relations in the sets $\mathcal{T}_\tau$. Due to Restriction (R3), this graph is acyclic, which is important as otherwise none of the tasks in a cycle could ever start due to a circular wait condition. Similarly, in the case of a mutual dependency between two or more tasks violating Restriction (R3), Algorithm 2 would run in an infinite recursion as it calls itself with the same inputs infinitely often.

Like with most scheduling problems, the production of different products, i.e., different jobs, is considered to be independent. This is the case as the tasks in different jobs are independent from one another as specified in Restriction (R2). In line with Nasiri and Kianfar [60], we consider the interdependence between different jobs to be rare in practice. In our setup they only arise as the jobs are manufactured in the same production facility and share the same machine pool.

To summarize, precedence relations are only present between tasks in the same job. For the precedence relations in a job, all relations that can be represented by an arbitrary acyclic graph can be considered in the scheduling problem, as it is the case for some closely related scheduling problems [6, 60].

**Example 1** (Precedence graph). *Consider a job $J$ that consists of the tasks $\mathcal{T}_J = \{\tau_1, \ldots, \tau_7\}$. Let the precedence relations between the tasks be given as follows:*

$$\mathcal{T}_{\tau_1} = \mathcal{T}_{\tau_2} = \mathcal{T}_{\tau_3} = \emptyset \qquad \mathcal{T}_{\tau_4} = \{\tau_1\} \qquad \mathcal{T}_{\tau_5} = \{\tau_3\}$$
$$\mathcal{T}_{\tau_6} = \{\tau_2, \tau_3, \tau_5\} \qquad \mathcal{T}_{\tau_7} = \{\tau_6\}$$

*The resulting precedence graph is depicted in Figure 3.2. Resolving the indirect dependencies of the tasks with Algorithm 2 leads to the sets $\bar{\mathcal{T}}_{\tau_i} = \mathcal{T}_{\tau_i}$ for $i = 1, \ldots, 6$ and the set $\bar{\mathcal{T}}_{\tau_7} = \{\tau_2, \tau_3, \tau_5, \tau_6\}$ for task $\tau_7$.*

Example 1 shows the different possible relations between tasks in the same job. First of all, the relation between the tasks $\tau_1$ and $\tau_4$ and the remaining tasks shows that no relation between all tasks in the same job is required and the precedence graph does not need to be connected. The dependency of task $\tau_6$ on three tasks, which partly depend on one another themselves, illustrates the arbitrary relations between the tasks. One task can have multiple predecessors, as it is the case for $\tau_6$, and multiple successors, as it is the case for $\tau_3$. The dependency of task $\tau_6$ on $\tau_3$ could be removed in the interest of sparsity of the representation since $\tau_6$ can not be started before the task $\tau_5$ is finished, which itself requires $\tau_3$ to be completed. However, such a redundancy in the precedence relations does not cause any issues with the methods introduced in the remainder of this thesis. The dependency of $\tau_7$ on $\tau_6$ and the resulting indirect dependency on the tasks $\tau_2, \tau_3$ and $\tau_5$ illustrates the relation between the sets $\mathcal{T}_{\tau_7}$ and $\bar{\mathcal{T}}_{\tau_7}$.

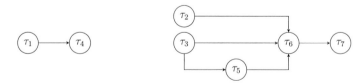

Figure 3.2: Precedence graph of job $J$ in Example 1. The arrows illustrate that the task at its beginning enables the task at its end.

Not only the tasks but also the machines and jobs are restricted in the real-world production process. Every machine has a limited production capacity, which we assume to be one for simplicity, and we assume that every machine has an infinite output buffer. For the jobs, we assume that they represent products that can only be at one place, and thus they can only be handled by a single machine at a time. Together with the limited capabilities of the machines, the following restrictions are present, which are all the mostly considered case in the literature on scheduling of FJSs [13, 15, 24, 80]:

(R5) Every machine $M$ can only execute a subset of tasks, which is specified as $\mathcal{T}_M \subset \mathcal{T}$.

(R6) A machine can only execute one task at a time.

(R7) A job can only be handled by one machine at a time. With Restriction (R6), this implies that only one task per job can be executed at a time.

Despite these restrictions, which are rather few compared with other scheduling problems in manufacturing, the production problem still has a lot of flexibility as introduced in Section 2.2 to classify the JMPMSF with respect to other scheduling problems. The JMPMSF offers the five types of flexibility (F1) - (F5), which can be exploited to react to changes in the production process. The Flexibility (F3), meaning the possibility to change the sequence in which the different tasks in the same job are executed, is restricted by Restriction (R1). This restriction, however, is required to formulate realistic production scenarios, which can be seen at countless examples. An obvious one is that it is not reasonable to package the product, before it has been produced. Therefore, the packaging task must be performed after the tasks corresponding to the actual production.

Most scheduling problems considered more frequently in literature, as for example the job shop (JS), the job shop with multi-murpose machines (JMPM) and the partial job shop (PJS) have further restrictions [10, 66]. In the JMPM, every job has a predefined process plan in the form of a fixed sequence in which the tasks in the job have to be executed, i.e., it does not offer sequence flexibility (F3). In the PJS, every task can only be executed by one distinct machine, i.e., it does not offer machine flexibility (F1). The JS combines both of these restrictions. The more restrictive cases can be considered as special cases of the JMPMSF in the presented setup. First, the restriction to a predefined process plan as for the JS and the JMPM can be formulated through a special linear precedence graph and the sets $\mathcal{T}_\tau$ in Restriction (R1). The restriction of the machines to only have one specific capability, as it is the case in the JS and the PJS, results in sets $\mathcal{T}_M$ with only one element in Restriction (R5). In addition, other even more restrictive scheduling problems can be considered as special cases of the JMPMSF. Those include the open shop and the flow shop, to only name the most prominent ones. We refer to [10] and [65] for their definition.

The benefit of having a problem description with different kinds of flexibility is that they offer the possibility to find an improved schedule with respect to one that was determined for a nominal case which is not guaranteed to persist over time. Determining an improved schedule might be useful to increase the performance of the system and in some cases it is even required to achieve the desired production goal despite unplanned changes in the manufacturing system as for example machine dropouts. The downside of having a lot of flexibility is that it leads to a high complexity of the scheduling problem and it is hard to find an optimal solution [24, 40]. In fact, the considered problem is NP-hard as it is a generalization of the flexible job shop (FJS), as discussed in Section 2.2, and since the FJS is NP-hard itself [6, 15, 24]. Increasing the flexibility of the scheduling problem with respect to the FJS makes a hard to solve problem even harder due to the larger solution space. To exploit the provided flexibility in order to achieve a desired goal, we discuss the scheduling objective in the JMPMSF in the next section.

### 3.2.3 Scheduling Objective

The goal of the proposed method is to optimize the throughput of the manufacturing system in order to maximize its profitability. This means that the best possible assignment between operations and machines needs to be found. Moreover, as the system evolves over time as the production proceeds, also the timing of the assignment needs to be considered. Hence, the goal is to find the best possible schedule of jobs in the manufacturing system with a given set of machines. As defined in Section 2.2 and in line with the notion of Birgin et al. [6], a production schedule is the assignment of the operations to the machines and a starting time. A schedule is considered to be feasible if the restrictions (R1) and (R4)–(R7) are met. Restrictions (R2) and (R3) refer to the formulation of the jobs and therefore do not need to be explicitly considered during scheduling.

As in real production scenarios, where not all future jobs are known and new jobs may arise at arbitrary points in time, we also consider that jobs may arrive during runtime. Therefore, the goal of the proposed scheme is not necessarily to find *the best* schedule for the scheduling problem as it is known at some particular time instant. We rather propose a *scheduling policy*, as, for example, introduced by Pinedo [65], which is able to generate feasible schedules in different states of the manufacturing system that are as good as possible under the given circumstances. In order to arrive at a schedule that is "as good as possible", meaning that it is desirable from a manufacturing point of view, the profit needs to be quantified with a cost function. In general, a cost function has to take into account the criteria that matter in the specific production scenario with respect to the economic goals of the production system. This might for example be the usage or waste of material or energy, the storage cost, or even the environmental impact as proposed in [69]. The design of the cost function offers the possibility to set priorities according to quantifiable criteria. Single objectives as the minimization of the makespan, total flow time, and many more are frequently considered in academic scenarios. As usually multiple criteria need to be considered to capture the true manufacturing cost and reward, multi-objective scheduling problems address real-world scenarios in a better way [80]. In such problems, the computation of comparable cost values for different possibilities can be determined with a weighted evaluation as suggested in [69]. As our focus is not to design meaningful scheduling problems but to develop a general scheme to solve them, we assume a meaningful cost function to be given. In order to guarantee the desired properties of the scheduling scheme, the cost function needs to fulfill a specific criterion, which will be discussed in Chapter 4. In Section 4.4.2, we discuss how this criterion relates to the cost of different elements in the problem formulation and thereby illustrate how the economic performance criteria for the manufacturing system can be considered while preserving the theoretic guarantees of the scheduling scheme.

# 3.3 Automatic Petri Net Generation for the JMPMSF

For the problem described in Section 3.2, we generate a mixed integer program (MIP) model by means of an automated procedure described in Section 3.3.2. In contrast to existing MIP models for the JMPMSF [6, 51, 63], we introduce an intermediate step to tackle the complexity of the problem with a modular approach in terms of a Petri net (PN). The model we use was first described in [87]. It is based on the fundamental form of a PN introduced by Carl Adam Petri in his dissertation [64]. We exploit the possibility to use algebraic techniques and in particular integer programming to describe and optimize the dynamic behavior of a system modeled as PN, as suggested by Giua and Seatzu [25]. In Section 3.3.1, we slightly modify the fundamental PN formulation in order to distinguish controlled transitions from independent ones in order to increase the descriptive power of PNs. The controlled transitions represent the active means to influence the production system, which is required to exploit the flexibilities in the model for scheduling and control purposes as described in Section 3.2. In contrast to that, the independent transitions represent the autonomous evolution of the production process. The generation of the PN model from the formulation of the JMPMSF provided in Section 3.2 is introduced in Section 3.3.2. The model generation is done with specific algorithms in an automated procedure and results in a discrete time description of the production process. It is explained how the algorithms consider the restrictions presented in Section 3.2.2 and how they enable the required flexibilities described therein. In Section 3.3.3, we analyze the resulting PN and discuss its properties. First, we justify that the PN representation is suited for the intended purpose before we provide important features that can be exploited for the scheduling scheme in Chapter 4. The present section is based on [87] and [88] and taken in parts literally from [88].

## 3.3.1 Modified Petri Net Formulation

In Section 2.3, we only introduced common notions for PNs that can be found in literature, for example in [11, 73]. However, in order to describe the scheduling problem introduced in Section 3.2, we will slightly adapt the role of the transitions. In most literature on PNs, the firing of transitions is initiated by predefined rules, considered to be stochastic, or the dynamics is even seen as the set of all possible sequences in which the transitions can fire [3, 66, 73]. Those representations are not suited to model the possibility to actively influence the system and exploit the available flexibility. Therefore, we will divide the set of transitions $\mathbb{T}$ and use a subset $\mathbb{T}_C \subset \mathbb{T}$ of the transitions as controlled inputs to actively influence the dynamics of the PN, which we will call *controlled part* of the PN. Another subset $\mathbb{T}_I \subset \mathbb{T}$ of the transitions is defined to fire as soon as they are enabled, which we will call *independent part* of the PN. As discussed by Giua and Seatzu [25], the PN dynamics (2.2) is a special case of the dynamics of a linear discrete time system

$$x(k + 1) = Ax(k) + Bu(k) \tag{3.1}$$

with $A = I$, where $I$ is the identity matrix of appropriate dimension. We will now exploit the possibility to change the matrix $A$ such that it reflects the influence of the *independent transitions* in the set $\mathbb{T}_I$. The matrix $B$ then merely considers the actively *controlled transitions* from the set $\mathbb{T}_C$ and the control decisions are captured in the firing count vector $u$.

In order to determine the matrix $A$, we need to discuss under which circumstances it is possible to represent the firing of some transition $T$ "as soon as it is enabled" by a multiplication of the state with a constant matrix without violating the constraint (2.3). Only if this is possible, one can express the firing of an independent transition $T$ by means of a matrix multiplication $Ax$ of the PN's state $x$ with a matrix $A$. Whether this is the case depends on the interconnection of the transition $T$ with the rest of the PN. Therefore, we introduce three restrictions on the independent transitions. A transition $T$ can only be considered to be independent, i.e., $T \in \mathbb{T}_I$, if

(I1) it only has one input place $P_i$,

(I2) its only input place $P_i$ has no second output transition,

(I3) the weight of the arc $(P_i, T)$ connecting the place $P_i$ to its output transition $T$ must be $w(P_i, T) = 1$.

With those restrictions, only the basic PN elements depicted in Figure 3.3, i.e., *sequence*, *merging*, and *splitting*, can be assigned to the independent part of the PN, i.e., to the set $\mathbb{T}_I$. On the other hand, *conflict* and *synchronization* depicted in Figure 3.4 *must* be controlled actively and the corresponding transitions always belong to the set $\mathbb{T}_C$. This obviously also holds for the combination of synchronization and conflict in a so called *non free-choice conflict* [3].

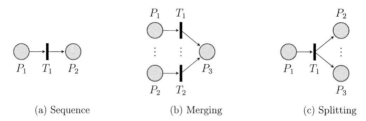

| (a) Sequence | (b) Merging | (c) Splitting |

Figure 3.3: Elementary Petri net structures in which the transitions can be assigned to the set of independent transitions $\mathbb{T}_I$.

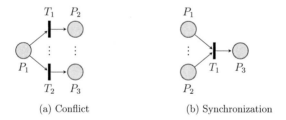

| (a) Conflict | (b) Synchronization |

Figure 3.4: Elementary Petri net structures in which the transitions must be actively controlled and always belong to the set of controlled transitions $\mathbb{T}_C$.

With the distinction between independent and controlled transitions, the PN dynamics is described as a linear time invariant system (3.1) with the linear inequality constraint (2.3) that ensures that the places do not hold a negative number of tokens

$$x(k+1) = Ax(k) + Bu(k), \qquad x(0) = x^0 \tag{3.2a}$$

$$0 \le x(k) - B^- u(k). \tag{3.2b}$$

As there always has to be an integer number of tokens, the newly introduced dynamics matrix must be a matrix of natural numbers, i.e., $A \in \mathbb{N}^{n \times n}$. The dynamics (3.2a) together with the non-negativity constraint (3.2b) is called *state space description* or *state space form* of the PN.

If a transition $T$ is actively controlled, i.e., $T \in \mathbb{T}_C$, its effect is captured in the input matrix $B$ composed of the matrices $B^+$ and $B^-$ which are created as usual, as we describe in (2.1). To introduce the independent part of the PN, the matrix $A$ is changed starting from an identity matrix $I$ for a completely controlled PN. If the output transition $T_i$ of a place $P_i$ is handled as an independent transition, i.e., $T_i \in \mathbb{T}_I$, the $i$-th column of the matrix $A$ is changed. The column of the incidence matrix $B$ corresponding to transition $T_i$ is determined as described in Section 2.3 with (2.1), but not introduced as a column in the input matrix $B$, but added to the $i$-th column of $A$. By that, the input arc $(P_i, T_i)$ of transition $T_i$ sets the diagonal entry to zero, i.e., $A_{i,i} = 0$, and the output arcs $(T_i, P_o)$ of the transition $T_i$ introduce the weights $w(T_i, P_o)$ in the entries $A_{o,i}$, i.e., $A_{o,i} = w(T_i, P_o)$. If a place has no independent output transition, a one in the corresponding diagonal entry in $A$ keeps its number of tokens constant as long as no controlled transition fires.

**Example 2.** *The matrices corresponding to the elementary PN structures in Figure 3.3 are*

$$A_{(a)} = \begin{bmatrix} 0 & 0 \\ 1 & 1 \end{bmatrix}, \qquad A_{(b)} = \begin{bmatrix} 0 & 0 & 0 \\ 0 & 0 & 0 \\ 1 & 1 & 1 \end{bmatrix} \quad \text{and} \quad A_{(c)} = \begin{bmatrix} 0 & 0 & 0 \\ 1 & 1 & 0 \\ 1 & 0 & 1 \end{bmatrix}.$$

*The firing of the independent transitions at the state $x(k)$ removes all tokens from the input places through the zero in the corresponding diagonal entry of the matrix $A$. In the state $x(k+1)$, tokens are produced in the output places of the independent transitions through the non-zero off-diagonal elements in $A$.*

Note that the criteria (I1) - (I3) are necessary so that a transition can be assigned to the set $\mathbb{T}_I$. If one of them is not fulfilled for a specific transition, it must be assigned to the set $\mathbb{T}_C$. However, it is still possible to assign transitions to $\mathbb{T}_C$ if they fulfill the three criteria (I1) - (I3). For example, it is possible to set the transition $T_2$ in Figure 3.3b as controlled transition and assign it to $\mathbb{T}_C$. In this case, the dynamics of the PN in Figure 3.3b is

$$x(k+1) = \begin{bmatrix} 0 & 0 & 0 \\ 0 & 1 & 0 \\ 1 & 0 & 1 \end{bmatrix} x(k) + \begin{bmatrix} 0 \\ -1 \\ 1 \end{bmatrix} u(k).$$

We will exploit the independent transitions to have a production process running autonomously while the scheduling decisions are handled by means of controlled transitions. How the problem formulation in Section 3.2 can be automatically transformed into a PN will be discussed in the next section.

### 3.3.2 Automatic Generation of the Discrete Time Petri Net Model

From the description of the JMPMSF in Section 3.2, we automatically generate a PN model with the algorithms presented in [87]. The generated PN model will provide all possible manufacturing decisions in the form of controlled transitions in the set $\mathbb{T}_C$. Their implications are captured in the matrix $B$ of the algebraic description (3.2a) and the decisions are taken by means of the firing count vector $u$. The automatic production process is represented through the independent transitions in the set $\mathbb{T}_I$ and handled in the matrix $A$ of the algebraic description (3.2a), as presented in Section 3.3.1. The state $x$ of the PN is the current status of the manufacturing system.

In order to preserve the production related meaning of the elements of the PN, we introduce distinct identifiers as labels for the places and transitions. By introducing corresponding vectors $x^{\text{ID}}$ and $u^{\text{ID}}$, the state $x$ and the input vector $u$ are intuitively understandable and can be related to the elements in the production system. The identifiers hold information on the machine, the task and the job to which the places and transitions refer. Additionally, we introduce a classifying character $\sigma \in \Sigma = \{S, F, P_q, B, I, N, C\}$ to indicate the meaning of the places and transitions in the manufacturing process. The classifier S relates to "start", F relates to "finish", B to "buffer", I to "idle", N to "necessary", and C to "completed". Moreover, multiple classifiers $P_q$ are used to indicate the progress of the production process by means of the index $q$. This set of classifiers was specifically selected for the given production problem together with the desired model of it and the algorithms that translate between them, which will be presented in the following. The set of classifiers is certainly not the only possible one to implement a similar model generation. If further characteristics of the production problem are investigated or other restrictions are present, the set of classifiers can be changed and further PN structures can be introduced.

The identifier of a place $P$ has the form $(M, \tau, J, \sigma)$, as a place can relate to one machine $M \in \mathcal{M}$, one task $\tau \in \mathcal{T}$, one job $J \in \mathcal{J}$ and has the production related meaning that corresponds to its classifier $\sigma \in \Sigma$. With this notation it becomes obvious to which machine, which task and which job a place $P_{(M,\tau,J,\sigma)}$ relates and how its marking can be interpreted by means of adjusted algorithms. If a place has no specific machine, task or job it relates to, its identifier has a zero at the respective position. For example the marking of the place $P_{(M,0,0,\mathrm{I})}$ indicates whether the machine $M$ is idle, which is independent of any specific task or job. The identifiers of the places correspond to the entries of the vector $x^{\mathrm{ID}}$ and connect the number of tokens in a place to an entry in the state vector $x$. The number of tokens in a place $P_{(M,\tau,J,\sigma)}$ at a time instant $k$ is represented by the entry $x_i(k)$ in the state vector $x(k)$ at the position $i$ where the corresponding entry in the vector of identifiers is $x_i^{\mathrm{ID}} = (M, \tau, J, \sigma)$.

A transition connects at least two places and its production related role can be related to the places connected to it. Its identifier has the form $(M, M', \tau, \tau', J, \sigma)$, which indicates that the transition $T_{(M,M',\tau,\tau',J,\sigma)}$ connects places of the form $P_{(M,\tau,J,\sigma)}$, and $P_{(M',\tau',J,\sigma)}$. The identifiers of transitions only refer to one job $J$, since we assume dependencies between different jobs to arise only through the fact that they are produced in the same manufacturing system, as discussed in Section 3.2 and formulated in Restriction (R2). As a consequence, all places connected to the same transition either relate to the same job or they do not relate to any job at all. The identifiers of the transitions correspond to the entries of the vector $u^{\mathrm{ID}}$ and connect the number of firings of a transition to an entry in the firing count vector $u$. The entry $u_i(k)$ in the firing count vector $u(k)$ at the position $i$ where $u_i^{\mathrm{ID}} = (M, M', \tau, \tau', J, \sigma)$ counts the number of firings of the transition $T_{(M,M',\tau,\tau',J,\sigma)}$ at the time instant $k$.

The resulting PN model is a discrete time description of the manufacturing system and the production processes are represented by chains of places $P_{(M,\tau,J,\mathrm{P}_1)}$, $P_{(M,\tau,J,\mathrm{P}_2)}, \ldots, P_{(M,\tau,J,\mathrm{P}_{k_\mathrm{P}(M,\tau)})}$. In the literature on FJS scheduling, such a model is also called "time-indexed" model. In order to arrive at a discrete number of production steps, the production time $\mathsf{t}_\mathrm{P}(M, \tau)$ is discretized. The number of production steps

$$k_\mathrm{P}(M, \tau) = \left\lceil \frac{\mathsf{t}_\mathrm{P}(M, \tau)}{\mathsf{t}_s} \right\rceil \qquad (3.3)$$

for the execution of the task $\tau$ on the machine $M$ is calculated with respect to the *sampling time* $\mathsf{t}_s$, which is also called *sampling period*. It is rounded to the next larger integer which means that for a large sampling time $\mathsf{t}_s$ the resulting production time $\bar{\mathsf{t}}_\mathrm{P} = k_\mathrm{P}\mathsf{t}_s$ might be larger than the true production time $\mathsf{t}_\mathrm{P}$, i.e., $\bar{\mathsf{t}}_\mathrm{P} \geq \mathsf{t}_\mathrm{P}$.

For the PN model of the manufacturing system, Algorithm 3 generates the set of places $\mathbb{P}$ for the PN, before Algorithm 4 completes the PN by generating the corresponding transitions. First, in Lines 1–3 of Algorithm 3, the *idle places* $P_{(M,0,0,\mathrm{I})}$, which indicate whether a machine is working or not, are created for all machines $M \in \mathcal{M}$. As it is assumed that all machines are idle at initialization, the idle places are marked with one token. This token will be consumed by every transition starting the execution of a task on the respective machine and returned once the production is finished. Due to this mechanism it is guaranteed that every machine only executes one task at a time and therefore respects Restriction (R6).

In Lines 4–14, the places reflecting the status of each job $J \in \mathcal{J}$ are created. Every job $J$ has a *starting place* $P_{(0,0,J,\mathrm{S})}$, initialized in Line 5 with one token, indicating that the production of the respective job has not yet been started.

---

**Algorithm 3:** Create Places [87].

---

**input** : A set of machines $\mathcal{M}$, a set of tasks $\mathcal{T}$, a set of jobs $\mathcal{J}$

**output:** A set of places $\mathbb{P}$ marked with the initial state $x^0$

---

1  **for** *every machine* $M \in \mathcal{M}$ **do**

2  |  Add an idle place $P_{(M,0,0,\mathrm{I})}$ to $\mathbb{P}$ holding one token;

3  **end**

4  **for** *every job* $J \in \mathcal{J}$ **do**

5  |  Add the starting place $P_{(0,0,J,\mathrm{S})}$ to $\mathbb{P}$ holding one token;

6  |  **for** *every task* $\tau \in \mathcal{T}_J$ **do**

7  |  |  Add an unmarked completion place $P_{(0,\tau,J,\mathrm{C})}$ and a necessity place $P_{(0,\tau,J,\mathrm{N})}$ holding one token to $\mathbb{P}$;

8  |  |  **for** *every machine* $M \in \mathcal{M}$ **do**

9  |  |  |  **if** *task* $\tau \in \mathcal{T}_M$ **then**

10 |  |  |  |  Add $k_{\mathrm{P}}(M,\tau)$ unmarked production places $P_{(M,\tau,J,\mathrm{P}_1)}, \ldots, P_{(M,\tau,J,\mathrm{P}_{k_{\mathrm{P}}(M,\tau)})}$ and an unmarked buffer place $P_{(M,\tau,J,\mathrm{B})}$ to $\mathbb{P}$;

11 |  |  |  **end**

12 |  |  **end**

13 |  **end**

14 **end**

---

For every task $\tau \in \mathcal{T}_J$ of every job $J$, multiple production places and buffer places are created in Line 10. For all machines $M$ with $\tau \in \mathcal{T}_M$, the *production places* $P_{(M,\tau,J,\mathrm{P}_q)}$, $q = 1, \ldots, k_\mathrm{P}(M,\tau)$ are generated. They indicate at which machine $M$ the task $\tau$ of job $J$ is executed and how far its execution already proceeded. At the end of Line 10, the *buffer places* $P_{(M,\tau,J,\mathrm{B})}$ are created. If a buffer place $P_{(M,\tau,J,\mathrm{B})}$ is marked, this shows that the task $\tau$ of the job $J$ was processed on machine $M$ at last and that the semi-finished product of this job is currently in the output buffer of machine $M$. Once the production of a job starts, the token that is initialized in the starting place $P_{(0,0,J,\mathrm{S})}$ is moved through the production places $P_{(M,\tau,J,\mathrm{P}_q)}$ and buffer places $P_{(M,\tau,J,\mathrm{B})}$ according to the production process. The machine at which the job is being processed or was processed last can be deduced from the identifiers and the markings of the production and buffer places related to it. By generating the production and buffer places for all possible machines that can execute some task, the selection of the machine that will eventually execute it is not predefined a priori and can be chosen during runtime. This structure enables the Flexibility (F1).

In Line 7, *necessity places* $P_{(0,\tau,J,\mathrm{N})}$ and *completion places* $P_{(0,\tau,J,\mathrm{C})}$ are created for all tasks $\tau \in \mathcal{T}_J$, which indicate the production progress of job $J$. Besides that, the completion places are used to ensure through a mechanism generated in Algorithm 4 that the precedence relations between the tasks are met and Restriction (R1) is respected. As it is assumed that no task in any job is already completed at initialization, the completion places are initialized unmarked whereas the necessity places hold one token indicating that the respective task needs to be executed once.

In Algorithm 4, the transitions and arcs that allow the possible evolutions of the system and implement the required constraints are generated. At first, in Lines 5 and 6, the starting transitions $T_{(0,M,0,\tau,J,\mathrm{S})}$ to start the jobs are created for every possible *initial* task of the jobs. A task $\tau$ is a possible initial task of a job if it does not depend on any other task that needs to be completed before it can be started, i.e., $\mathcal{T}_\tau = \emptyset$.

The remaining starting transitions $T_{(M,M',\tau,\tau',J,\mathrm{S})}$ allowing to start all further tasks in various different sequences are generated in Lines 15 – 17. The possible sequences in which the different tasks in a job can be started are only restricted by the precedence between the tasks, i.e., Restriction (R1). The precedence relations are considered in the if-statement in Line 13 of Algorithm 4. It makes sure that a starting transition $T_{(M,M',\tau,\tau',J,\mathrm{S})}$ which starts the task $\tau'$ directly after another task $\tau$ is only created if $\tau'$ can be a direct successor of $\tau$. The first two conditions in this if-statement make sure that the tasks $\tau$ and $\tau'$ are different from one another and that $\tau'$ must not have to be completed before $\tau$ can be started. The third condition ensures that there is no third task $\tau^*$ which has to be executed "between" $\tau$ and $\tau'$, meaning that it has to be completed before $\tau'$ can be started, but cannot be started itself before $\tau$ was completed.

Every starting transition is a synchronization of an idle place, a necessity place, a starting or buffer place, and possibly multiple completion places as illustrated in Figure 3.5. Its input places need to be marked before a starting transition is enabled and allowed to fire, which is why it has to be implemented as a controlled transitions.

---

**Algorithm 4:** Create Transitions and Arcs [87].

---

**input** : A set of machines $\mathcal{M}$, a set of tasks $\mathcal{T}$, a set of jobs $\mathcal{J}$, a set of places $\mathbb{P}$

**output:** A set of independent transitions $\mathbb{T}_I$, a set of controlled transitions $\mathbb{T}_C$, a set of arcs $\mathbb{E}$

1 **for** *every job $J \in \mathcal{J}$* **do**
2    **for** *every task $\tau \in \mathcal{T}_J$* **do**
3      **for** *every machine $M$ with $\tau \in \mathcal{T}_M$* **do**
4        **if** $\mathcal{T}_\tau = \emptyset$ **then**
5          Add a starting transition $T_{(0,M,0,\tau,J,S)}$ to $\mathbb{T}_C$;
6          Add its input arcs $(P_{(0,\tau,J,N)}, T_{(0,M,0,\tau,J,S)})$, $(P_{(M,0,0,I)}, T_{(0,M,0,\tau,J,S)})$ and $(P_{(0,0,J,S)}, T_{(0,M,0,\tau,J,S)})$
         and its output arc $(T_{(0,M,0,\tau,J,S)}, P_{(M,\tau,J,P_1)})$ to $\mathbb{E}$;
7        **end**
8        Add the production transitions $T_{(M,M,\tau,\tau,J,P_q)}$, $q \in [1, k_P(M, \tau) - 1]$ to $\mathbb{T}_I$;
9        Add their input arcs $(P_{(M,\tau,J,P_q)}, T_{(M,M,\tau,\tau,J,P_q)})$
       and their output arcs $(T_{(M,M,\tau,\tau,J,P_q)}, P_{(M,\tau,J,P_{q+1})})$ to $\mathbb{E}$;
10        Add a finishing transition $T_{(M,M,\tau,\tau,J,F)}$ to $\mathbb{T}_I$;
11        Add its input arc $(P_{(M,\tau,J,P_{k_P(M,\tau)})}, T_{(M,M,\tau,\tau,J,F)})$
       and its output arcs $(T_{(M,M,\tau,\tau,J,F)}, P_{(M,\tau,J,B)})$, $(T_{(M,M,\tau,\tau,J,F)}, P_{(M,0,0,I)})$ and $(T_{(M,M,\tau,\tau,J,F)}, P_{(0,\tau,J,C)})$ to $\mathbb{E}$;
12        **for** *every task $\tau' \in \mathcal{T}_J$* **do**
13          **if** $\tau \neq \tau'$ **and** $\tau' \notin \bar{\mathcal{T}}_\tau$ **and** $\{\tau^* \in \mathcal{T}_J : \tau \in \bar{\mathcal{T}}_{\tau^*}, \tau^* \in \bar{\mathcal{T}}_{\tau'}\} = \emptyset$ **then**
14            **for** *every machine $M'$ with $\tau' \in \mathcal{T}_{M'}$* **do**
15              Add a starting transition $T_{(M,M',\tau,\tau',J,S)}$ to $\mathbb{T}_C$;
16              Add its input arcs $(P_{(0,\tau',J,N)}, T_{(M,M',\tau,\tau',J,S)})$, $(P_{(M',0,0,I)}, T_{(M,M',\tau,\tau',J,S)})$ and $(P_{(M,\tau,J,B)}, T_{(M,M',\tau,\tau',J,S)})$
             and its output arc $(T_{(M,M',\tau,\tau',J,S)}, P_{(M',\tau',J,P_1)})$ to $\mathbb{E}$;
             **for** *every task $\tau'' \in \mathcal{T}_{\tau'}$* **do**
17                 Add the arcs $(P_{(0,\tau'',J,C)}, T_{(M,M',\tau,\tau',J,S)})$ and $(T_{(M,M',\tau,\tau',J,S)}, P_{(0,\tau'',J,C)})$ to $\mathbb{E}$;
18              **end**
19            **end**
20          **end**
21        **end**
22      **end**
23    **end**
24 **end**

---

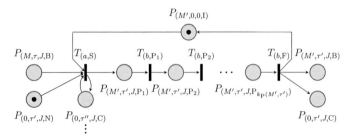

Figure 3.5: Illustration of a production sequences for the execution of the task $\tau'$ of the job $J$ on machine $M'$ as it is generated in the Algorithms 3 and 4, with $a = (M, M', \tau, \tau', J)$ and $b = (M', M', \tau, \tau', J)$. The tokens in the sequence are depicted as they are initialized in Algorithm 3. As at initialization no task $\tau$ or $\tau''$ is already completed, the places $P_{(M,\tau,J,B)}$ and $P_{(0,\tau'',J,C)}$ are not marked and the transition $T_{(M,M',\tau,\tau',J,S)}$ is not enabled.

This is not a restriction, however, as actively choosing the sequence in which the different tasks of the different jobs are started on the available machines by deciding which starting transitions fire at a given time exactly implements the desired Flexibilities (F3) and (F4). The starting of a task reserves the machine $M$ that is used for its execution by consuming the token from the idle place $P_{(M,0,0,I)}$. This guarantees that every machine is only used once at a time as required by Restriction (R6). By consuming the token from the necessity place $P_{(0,\tau,J,N)}$, it is indicated that the task $\tau$ will not be necessary any longer. The consumption of the token from the starting place $P_{(0,0,J,S)}$ or the buffer place $P_{(M,\tau,J,B)}$ reserves the semi-finished product of the job $J$ and guarantees that only one task of every job is executed at a time, as required by Restriction (R7). Each starting transition that starts the execution of a task $\tau'$ consumes the token from all completion places $P_{(0,\tau'',J,C)}$ of the tasks $\tau'' \in \mathcal{T}_{\tau'}$ that need to be finished before the task $\tau'$ can be started. Thereby it is guaranteed the task $\tau'$ can only be started once its precedence constraints according to Restriction (R1) are fulfilled. The tokens are immediately returned to the completion places such that their production status is preserved. Starting transitions create a token in the first production place $P_{(M,\tau,J,P_1)}$, indicating that the production starts.

The production transitions $T_{(M,M,\tau,\tau,J,P_q)}$, which are created in the Lines 8 and 9 of Algorithm 4, implement the production process as a sequence of production steps similar to Figure 3.3a, as illustrated in Figure 3.5. They move tokens from one production place $P_{(M,\tau,J,P_q)}$, $q \in [1, k_P(M,\tau) - 1]$ to the next production place $P_{(M,\tau,J,P_{q+1})}$ and are implemented as independent transitions.

The finishing transitions $T_{(M,M,\tau,\tau,J,F)}$ are created in Lines 10 and 11 of Algorithm 4 and have the form of a splitting similar to Figure 3.3c, as illustrated in Figure 3.5. They are considered to be independent as well. A finishing transition consumes the token from the last production place $P_{(M,\tau,J,P_{k_p(M,\tau)})}$, returns a token to the idle place $P_{(M,0,0,I)}$ of the machine $M$ that finishes the production and stores the semi-finished product in its output buffer by generating a token in the place $P_{(M,\tau,J,B)}$. Finally, the finishing transition marks the task $\tau$ as completed by creating a token in the completion place $P_{(0,\tau,J,C)}$.

### State Space Description of the Petri Net

The state space description of the PN in the form of equations (3.2) is created as introduced in Section 3.3.1. The identification of the entries in the initial state $x^0$ and the matrices $A$, $B^+$ and $B^-$ corresponding to the places $P_{(M,\tau,J,\sigma)}$ and the transitions $T_{(M,M',\tau,\tau',J,\sigma)}$ are determined by means of the vectors of identifiers $x^{\mathrm{ID}}$ and $u^{\mathrm{ID}}$, respectively. First, the initial state $x^0$ is initialized with the initial markings of all places $P \in \mathbb{P}$. The effects of the production and finishing transitions are represented in the matrix $A$. Starting from an identity matrix $I$, for every production transition $T_{(M,M,\tau,J,P_q)}$ the diagonal entry $A_{i,i}$ corresponding to its input place $P_{(M,\tau,J,P_q)}$ is set to zero. The index $i$ is determined through the vector of identifiers $x^{\mathrm{ID}}$, where the entry in the $i$-th position is $x_i^{\mathrm{ID}} = (M,\tau,J,P_q)$. To consider the output of the transition $T_{(M,M,\tau,J,P_q)}$, a one is introduced at the entry $A_{j,i}$ where the row $j$ is the position at which $x_j^{\mathrm{ID}} = (M,\tau,J,P_{q+1})$. The resulting matrix is similar to $A_{(a)}$ in Example 2. For every finishing transition $T_{(M,M,\tau,J,F)}$, the diagonal entry corresponding to its input place $P_{(M,\tau,J,P_{k_p(M,\tau)})}$ is set to zero and ones are introduced in the same column in the rows corresponding to its output places $P_{(M,\tau,J,B)}$, $P_{(M,0,0,I)}$, and $P_{(0,\tau,J,C)}$. The result is similar to matrix $A_{(c)}$ in Example 2.

For every starting transition $T_{(M,M',\tau,\tau',J,S)}$, a column is added to the matrix $B$ through the matrices $B^+$ and $B^-$ as defined in (2.1). The new columns in $B^+$ and $B^-$ are initialized with zeros. In $B^-$, a one is introduced in the lines corresponding to the input places $P_{(0,\tau',J,N)}$, $P_{(M',0,0,I)}$, and $P_{(M,\tau,J,B)}$. The precedence relations among different tasks, which are considered through arcs from the completion places of other tasks $\tau'' \in \mathcal{T}_{\tau'}$, lead to further ones in the lines corresponding to possibly multiple places of the form $P_{(0,\tau'',J,C)}$. In $B^+$, a one is introduced in the line corresponding to the first production place $P_{(M',J,J,P_1)}$ and further ones are introduced in the lines corresponding to the completion places $P_{(0,\tau'',J,C)}$ of the precedence relation, in order to keep their marking constant. Note that directly creating the matrix $B$ and omitting the matrices $B^+$ and $B^-$ would make the completion places obsolete and cancel out their effect. Additionally, the matrix $B^-$ is required to formulate the non-negativity constraint (3.2b). The input vector $u$ has one entry for every starting transition.

The result of the Algorithms 3 and 4 is the Petri net $PN = (\mathbb{P}, \mathbb{T}, \mathbb{E}, w, x^0)$ and its state space description given by the matrices $A, B^+, B^-$ and the initial state $x^0$. The state space description will be used for the scheduling scheme described in Chapter 4. The proposed algorithms ensure certain properties of the derived PN and its state space representation which will be discussed in the next section.

### 3.3.3 Analysis and Discussion of the Discrete Time Petri Net Model

Before using the PN model generated in Section 3.3.2 for the scheduling of the JMPMSF in Section 4, we will analyze its properties and the resulting state space description (3.2). To this end, we first introduce the concept of a steady state in Definition 2. Then, in Lemma 1, we provide some specific properties that will be exploited to prove important features of the proposed scheduling scheme in Section 4.

**Definition 2** (Steady state). *A steady state of a linear discrete time system $x(k+1) = Ax(k) + Bu(k)$ is a pair of state and input $(x^s, u^s)$ with $x^s \in \mathbb{N}^n$, $u^s \in \mathbb{N}^m$ and $x^s = Ax^s + Bu^s$.*

**Lemma 1** (Properties of the automatically generated Petri net). *For the Petri net $PN = (\mathbb{P}, \mathbb{T}, \mathbb{E}, w, x^0)$ generated with the Algorithms 3 and 4, the following properties hold:*

*(P1)* *There is no steady state $(x^s, u^s)$ with $u^s \neq 0$, $u^s \in \mathbb{N}^m$. We will therefore henceforth omit input $u^s$ in the steady state denotation and call a state vector $x^s$ with $x^s = Ax^s$ a steady state.*

*(P2)* *The following statements are equivalent:*
*In the state $x^s$, no production is taking place, i.e., no production place $P_{(...,P)}$ is marked. ⇔ The state vector $x^s$ is a steady state.*

*(P3)* *The initial state $x^0$ generated in Algorithm 3 is a steady state and $x^0 \in \mathbb{N}^n$.*

*(P4)* *All parts $\bar{x}, \hat{x} \in \mathbb{N}^n$ of a steady state $x^s$ are themselves steady states, i.e.,*
*$x^s = \bar{x} + \hat{x} = Ax^s$ ⇔ $\bar{x} = A\bar{x}$, $\hat{x} = A\hat{x}$.*

*(P5)* *The precondition of every firing $x \geq B^- u$, which needs to be satisfied according to the non-negativity condition (3.2b) in order to start a production process, depends on a steady state $x^s = B^- u$ for all inputs $u \in \mathbb{N}^m$, i.e., $B^- u = AB^- u$ for all $u \in \mathbb{N}^m$.*

*(P6)* *An operation $O = (\tau, J)$ which has been completed stays completed, i.e., once a completion place $P_{(...,C)}$ is marked it stays marked and once a necessity place $P_{(...,N)}$ is unmarked it stays unmarked.*

*(P7)* *If $x(0) \in \mathbb{N}^n$, then $x(k) = A^k x(0) \in \mathbb{N}^n$ for all $k \geq 0$ and condition (3.2b) is satisfied for $x(k)$ with $u(k) = 0$ for all $k \geq 0$.*

*(P8)* *The following two statements are true and equivalent:*
*Every state $x$ eventually enters a steady state $x^s$ if no controlled transition $T \in \mathbb{T}_C$ is fired. ⇔ There is a $\bar{k} \in \mathbb{N}_{>0}$, $\bar{k} \leq n$ such that for all $k \geq \bar{k}$, it holds that $A^k = A^{(k+1)}$ and therefore $x^s = A^n x$ is a steady state.*

(P9) If an operation $O = (\tau', J)$, i.e., a specific task $\tau'$ in a job $J$, can be started at a steady state $x^{s,1}$, it can also be started in all steady states $x^{s,2}$ reachable from $x^{s,1}$ through a legal firing sequence respecting (3.2b), or it was already completed in $x^{s,2}$.

A proof of the Lemma 1 can be found in Appendix A. The Properties (P1)–(P6) define important features of steady states, that are not only important for theoretical analysis, but also required for a reasonable representation of a production system. Property (P1) states that the system can only stay at rest if no production process is actively started. If, in contrast to Property (P2), a production is going on while the system is at rest, either the model is flawed or the production does not yield any products, which would not be favorable. Initializing the system at rest as stated in Property (P3) is a common way of modeling. If Property (P4) was not fulfilled, there would be a steady state only due to two or more ongoing processes that cancel each other out. This can be reasonable for some models, for example, if the states describe inventory levels. In our case, however, as the individual products are tracked, this would imply that one process is progressing on a product while another one is reversing its effect, which is not desired. The absence of Property (P5) would allow the start of a new production process to be only temporarily possible and after some time it would not be possible any more without any external causes. Such a system behavior might originate from decaying products or tools, which are not considered in the presented setup. Property (P6) describes a production process without disintegrating products.

The created PN does not violate the non-negativity constraint (3.2b) by the independent evolution represented through the matrix $A$ as Property (P7) states. This means that the independent part of the model only leads to states which have a reasonable real-world representation, which is what non-negative tokens indicate. In the described production system no endless production processes can be started, which is expressed through Property (P8). Property (P9) shows that there are no restrictions in the sequence Flexibility (F3) of the described JMPMSF, except the ones explicitly modeled through Restriction (R1).

The discrete time model introduced in Section 3.3.2 describes the evolution of the production process with respect to the sampling time $t_s$. A smaller sampling time will lead to an increased state dimension, which results from (3.3) and the creation of the production places in Line 10 of Algorithm 3. On the other hand, a larger sampling time might lead to a more significant over-approximation of the true production times $t_P$ by the discrete time model through the rounding in (3.3). As a consequence, the sampling time $t_s$ should be chosen in accordance with the dynamics of the represented processes. If the manufacturing system has processes with vastly different time scales, it might be hard to find a good compromise. However, despite the increasing state dimension, the number of input variables, which has an even greater influence on the runtime of the optimization problems presented in Section 4, is independent of the sampling time. This alleviates the drawbacks of having a small sampling time and a large state dimension.

The presented mechanism of creating a PN model for a scheduling problem is not unique. With similar algorithms, further mechanisms and properties of a manufacturing system can be represented, as, for example, output buffers of the machines with limited capacity, production capacities of the machines larger than one, or maintenance tasks that are represented by transitions which are not related to any job. While the model introduced here is a binary PN, in which the places are either marked with one token or not marked at all and the weight of all arcs is one, the introduction of such effects might lead to non-binary PN. Due to the binary nature of the PN generated with the Algorithms 3 and 4, also its state space description is binary, meaning the matrices $A, B^-$ and $B^+$ only have the entries 0 or 1. In a non-binary PN on the other hand, there can be multiple tokens per place and arc weights different than one. By this means not only further mechanisms, but also more complex constraints of the manufacturing system can be introduced. For example the total amount of power that can be consumed by the production system at any given time can be limited by the introduction of energy tokens. Such tokens would initially be stored in special power capacity place. The number of tokens in this place represents the remaining power capacity of the system at any point in time. Initialized with the total available power, the starting transitions consume the number of tokens from this place according to the amount of power required for the corresponding production processes. If the power capacity place does not hold enough tokens, the starting transition is not enabled and thus the production process cannot be started.

The automatically created PN and the resulting MIP model is not the smallest one that would be possible for the given JMPMSF. For example, the size of the state vector could be reduced by summarizing all the buffer places $P_{(M,\tau_i,J,\mathrm{B})}$ for the different tasks $\tau_i \in \mathcal{T}_J$ of the same job $J$ that can be executed at the same machine $M$ into one buffer place $P_{(M,0,J,\mathrm{B})}$. By that, however, the information which task in job $J$ was executed last is removed from the state of the PN.

**Restrictions and flexibilities in the Petri net model**

In the creation of the discrete time PN model, several structures and mechanisms are created to represent the characteristics of the job shop with multi-purpose machines and sequence flexibility (JMPMSF). Subsequently, we will briefly highlight the most significant ones and discuss simplifications of the PN model for special cases of the JMPMSF at the example of the ordinary job shop (JS) problem. In general, the dependencies between the different parts of the JMPMSF are captured in the PN graph whereas the flexibility lies in the choice of the transitions to fire.

Restriction (R1), which expresses that the sequence of tasks in a job is not completely free but restricted according to the properties of the manufacturing process, is introduced through the arcs introduced in Line 17 of Algorithm 4. A starting transition $T_{(M,M',\tau,\tau',J,\mathrm{S})}$ of a task $\tau'$ is only enabled if all tasks $\tau''$ in the set $\mathcal{T}_{\tau'}$ are completed and their completion places are marked.

The limitation to the non-preemptive case formulated in Restriction (R4) is respected through the sequence of production places and production transitions introduced in Line 10 of Algorithm 3 and Lines 8 - 11 of Algorithm 4. Since this sequence cannot be interrupted, preemption of a task is not possible.

The restriction that every machine can only execute a subset of tasks, i.e., Restriction (R5), is already integrated in the problem formulation of the JMPMSF. It is integrated in the PN through checking the capabilities of the machines in the if-statement in Line 9 of Algorithm 3 before creating the production places corresponding to an investigated machine.

Restrictions (R6) and (R7), saying that each machine can only execute one task at a time and that only one task of every job can be executed at time, respectively, are considered by only introducing one token in the idle place of each machine in Line 2 of Algorithm 3 and in the starting place of each job in Line 5 of Algorithm 3. Through the arcs connecting the starting transitions to the idle places in Lines 6 and 16 of Algorithm 4, the token corresponding to a machine can only be assigned to one production process at a time. Similarly, the token corresponding to each job is moved from the starting place of the job through the production places to the buffer places of the machines. Thereby the semi-finished product of the job can only be at one place at a time, which realizes the Restriction (R7).

The Restrictions (R2) and (R3) constrain the possibilities of the problem formulation and therefore do not need to be represented in the PN.

The flexibilities of the JMPMSF are represented through the different possibilities to influence the PN model and decide about its evolution. The possibility to execute the same task on different machines, i.e., Flexibility (F1), is introduced through the different production places and buffer places for the same task, which are created in Line 10 of Algorithm 3. For an ordinary JS, the if-statement in Line 9 is only true for a single machine leading to a single chain of production places and a single buffer place for every task of every job.

The assumption of having multi-purpose machines in the JMPMSF, i.e., Flexibility (F2), is considered in the for-loop in Line 8 of Algorithm 3 together with the if-statement in Line 9, which allows to restrict the possibilities of the machines. For single-purpose machines in a JS, those conditions represent the search for the single machine that is able to execute the currently investigated task and could be implemented more efficiently.

The sequence flexibility in the JMPMSF, i.e., Flexibility (F3), is respected through the multitude of starting transitions that are created in the for-loop which starts in Line 12 of Algorithm 4. They allow to change the sequence in which the tasks $\tau \in \mathcal{T}_J$ in a job $J$ are executed. If no precedence relation between the tasks $\tau$ and $\tau'$ is present, both transitions $T_{(M,M',\tau,\tau',J,S)}$, starting $\tau'$ after $\tau$, and $T_{(M',M,\tau',\tau,J,S)}$, starting $\tau$ after $\tau'$, are present in the PN. It is guaranteed that only one of them can fire since each of them requires a token at the buffer place after the execution of the respective other task. For a JS without sequence flexibility, the if-statement in Line 13 of Algorithm 4 would only be true for the single task $\tau'$ that follows after the task $\tau$, leading to a single starting transition which starts $\tau'$ and recovering the given production sequence. Furthermore, the necessity places and completion places created in Line 7 of Algorithm 3 would be obsolete in this special case.

The Flexibility (F4) to decide on the sequence in which different operations are executed on a machine is considered through assigning the starting transitions to the controlled part of the PN. By that, a scheduling scheme, which has the PN as basis for its decisions, can determine which operations to execute first on a particular machine and initiate this production process by firing the corresponding starting transition.

The possibility to consider different types of jobs in the manufacturing system, i.e., Flexibility (F5), is strengthened by the automatic generation of the model from the JMPMSF description. By that, the modeling effort required to include new and different types of jobs is reduced. The Flexibilities (F4) and (F5) are inherent for scheduling schemes and therefore equally valid for the JS.

A new job $\hat{J}$ can be integrated in the PN model by executing the Algorithms 3 and 4 only for the new job. In Algorithm 3 this means running the for-loop starting in Line 4 only for $\hat{J}$ and in Algorithm 4 the main for-loop starting in Line 1 needs to be executed for $\hat{J}$. Since the different jobs are independent from one another as required in Restriction (R2), the already existing system model does not need to be changed. This simple mechanism allows to integrate new jobs with little effort and implements Flexibility (F5). Similarly, the introduction of a new machine $\hat{M}$ can be introduced by running the Algorithms 3 and 4 and only considering the new machine in the respective loops. Due to the modular structure of the JMPMSF and of its PN model, the introduction of a new machine leaves the existing model unchanged and introduces new flexibility that can be exploited by the scheduling scheme.

For a framework that is intended to run for a long period of time, not only including new elements but also removing obsolete ones is necessary. In a manufacturing system, this means that completed jobs need to be identified and removed. In the PN model this can be done by means of the completion places $P_{(0,\tau,J,C)}$ of the tasks $\tau \in \mathcal{T}_J$ in a job $J$. Once all tasks in a job $J$ are marked as completed, this job is accomplished and does not need to be considered any further. It can be removed from the JMPMSF by removing the corresponding elements and sets, and from the PN model by determining all corresponding places and transitions through their identifiers and removing them. Since the PN was built up modularly with the Algorithms 3 and 4 and since jobs are independent from one another in the JMPMSF, the remaining PN model is still valid for the remaining production system. In a similar way, also disused machines can be removed from the system. In this case, however, special attention needs to be paid that the production problem is still feasible. It needs to be ensured that despite the removal of the machine, all tasks still have at least one machine that can execute them.

The simple means to integrate new jobs and new machines, and to remove completed jobs and disused machines implement the desired modularity of a skill-based engineering approach as described in Section 2.1 and make the presented approach suitable for Industry 4.0 scenarios.

# 3.4 Summary

In this chapter, we introduced the MPC framework for the scheduling of FMSs. It constitutes a holistic approach from the problem formulation to its solution, since it automates the generation of a mathematical model and the model based scheduling of the FMS.

In Section 3.1, we started by explaining the different parts of this framework and their interplay in an initial overview. In Section 3.2, we investigated the scheduling problem for FMSs on the level of the problem description. We presented the JMPMSF as a general form of scheduling problems for FMSs. The general nature of the problem encompassing multiple more specific scheduling problems has the advantage that all methods developed on its basis are applicable to the specific cases and therefore promise to be relevant for many applications. We provided a structured definition of the problem and, by identifying the machines as the persistent element in the FMS, we tailored the problem description to an application in Industry 4.0 scenarios. Motivated by the digital twins of the components in the FMS, the problem description was formulated in a modular way. It is assumed that the machines are cyber-physical production systems that can be described as modules in the sense of a skill-based engineering approach and have a virtual representation in the form of their digital twin which is synchronized in real-time. In the scheduling problem, the machine modules need to be assigned to the modules corresponding to the products to be produced. The flexibilities resulting from this description of the JMPMSF, which are of particular interest in Industry 4.0 as already introduced in Section 2.2, are explicitly laid out. The restrictions that are posed in the problem formulation and complement the flexibilities of the JMPMSF are explained and it is discussed how further restrictions lead to special cases that are more widely discussed, as for example the well-known job shop problem.

The modularity of the problem description is exploited for the automatic model generation in the form of a PN in Section 3.3. We introduced the Algorithms 3 and 4 to transform the very general problem formulated in Section 3.2 into a mathematical model in the form of a PN. To this end, we enhanced the theory of PNs by separating an automatically running independent part from an actively controlled part. This does not change the graph description of the PN, but only the behavior of the transitions and thus the resulting algebraic description of the PN. The dependencies between the different parts in the problem description are captured in the PN graph whereas the flexibility lies in the choice of the transitions to fire. The result is a simple state space representation (3.2) in the form of a linear discrete time system with a linear inequality constraint. This algebraic description allows to apply methods from systems and control theory and specifically MPC as we will show in Chapter 4. In Section 5, we illustrate the applicability of the scheduling scheme with simulation examples from literature.

The Algorithms 3 and 4 are able to handle the JMPMSF in its generality and lead to simpler PN models for more specific special cases. The drawback of this generally applicable modeling approach is that it leads to rather high dimensional state space descriptions and therefore it is only suited for rather small problem instances. As discussed in Section 3.3.3, the sampling time offers the possibility to find a compromise between accuracy and dimension of the system. The flexibility and modularity of the problem description as well as the model generation is essential for its applicability in Industry 4.0 scenarios. The structural simplicity of the resulting mathematical model will be exploited in the formulation of the reactive scheduling scheme in the next chapter.

# Chapter 4

# Model Predictive Control for Flexible Job Shop Scheduling

On the basis of the state space description (3.2) of the automatically generated Petri net (PN), we will now develop two similar scheduling schemes for the job shop with multi-purpose machines and sequence flexibility (JMPMSF) based on model predictive control (MPC). Although they are specifically designed for JMPMSFs as they are defined in Section 3.2, they are exemplary for arbitrary scheduling problems formulated as a PN of the form introduced in Section 3.3.1.

We start by formulating the first MPC problem without terminal cost as described in [86], before we show that the problem is feasible in Section 4.1.1 and determine a criterion to guarantee the completion of a single operation in the production problem in Section 4.1.2. The insights form this investigation motivate the formulation of an MPC formulation with terminal cost in Section 4.2, where we show how also for this problem the completion of an operation can be ensured. The guarantees for the completion of the production problem for both MPC formulations are given in Section 4.3. In Section 4.4, we discuss some properties of the proposed MPC schemes. In particular, we investigate the robustness of the closed loop with respect to unforeseen changes in the system and the role of the cost function in the MPC problem with respect to the provided guarantees and the economic objective of the manufacturing system. We will briefly elaborate on the simplifications in the case of a linear cost function and alternative techniques to provide convergence guarantees of the MPC schemes. We end this chapter with a brief summary in Section 4.5. This section is based on [86] and [88] and taken in parts literally from [88].

## 4.1 MPC Formulation without Terminal Cost

As discussed in Section 3.2.3, the goal of the proposed production scheduling scheme is to optimize the profitability of the production system. This is done by assigning the different tasks to the most efficient machines and by choosing the production sequence which yields the most reward possible. In order to quantify the profitability of the production system represented as a PN, we introduce a cost function $c : \mathbb{X} \times \mathbb{U} \to \mathbb{R}$ that assigns to every pair $(x, u)$ of state and input a cost value $c(x, u)$. In particular, the cost function $c(x, u)$ quantifies how much cost the current status of the production system represented through the state $x$ and the decisions taken through the input $u$ cause. In general, the cost may depend arbitrarily on $x$ and $u$ and can also contain couplings between them, but we only consider the case of a linear cost function

$$c(x, u) = c_x^\top x + c_u^\top u \tag{4.1}$$

in more detail. As the state of the system changes over time and at every time step new decisions are taken, the cost $c(x(k), u(k))$ only represents the cost in the current time interval of length $t_s$. In order to achieve a good performance over time with respect to a given cost function $c$, the system needs to be influenced in a way such that the resulting cost summed up over time is minimal. In the sense of MPC, we formulate this goal as a finite horizon optimal control problem over the prediction horizon $N$, which is repeatedly solved in a receding horizon fashion [67]. It is initialized at each time instant $k$ with the most accurate knowledge about the system, which is the state $x(k)$ that we assume to be known or measured. This assumption is justified by the concept of the digital twin introduced in Section 2.1. The optimization problem to determine the MPC control law at time instant $k$ is formulated as

$$\underset{u(\cdot|k)}{\text{minimize}} \quad \sum_{\bar{k}=0}^{N-1} c\left(x\left(\bar{k}|k\right), u\left(\bar{k}|k\right)\right) \tag{4.2a}$$

$$\text{subject to} \quad x\left(\bar{k}+1|k\right) = Ax\left(\bar{k}|k\right) + Bu\left(\bar{k}|k\right), \tag{4.2b}$$

$$0 \leq x\left(\bar{k}|k\right) - B^- u\left(\bar{k}|k\right), \tag{4.2c}$$

$$u\left(\bar{k}|k\right) \in \mathbb{N}^m, \quad \text{for } \bar{k} = 0, \dots, N-1, \tag{4.2d}$$

$$x\left(0|k\right) = x\left(k\right). \tag{4.2e}$$

In the prediction, the PN dynamics (3.2a) and the non-negativity of the states (3.2b) are respected in the constraints (4.2b) and (4.2c). The optimal control problem is initialized with the measured state of the system $x(k)$ in the constraint (4.2e). The planned input vectors $u(\bar{k}|k)$ are only allowed to have non-negative integer values (4.2d), whereby also the predicted states $x(\bar{k}|k)$ remain vectors of non-negative integers. This can easily be verified by noticing that the initial state $x^0 \in \mathbb{N}^n$ and the matrices $A \in \mathbb{N}^{n \times n}$ and $B \in \mathbb{N}^{n \times m}$ only contain natural numbers and the non-negativity is imposed by constraint (4.2c). By non-negative inputs the tokens in the PN are moved in the intended direction. The optimization variables in the MPC problem are the inputs $u(\bar{k}|k)$ of the planned input trajectory $u(\cdot|k) = (x(0|k), \dots, u(N-1|k))$. They allow to choose a new firing count vector $u(\bar{k}|k)$ at every future time instant $k+\bar{k}$ and represent potential decisions taken for the production process. The minimizer of (4.2a) in the MPC problem is the optimal input trajectory $u^*(\cdot|k)$. The optimal predicted state trajectory is denoted $x^*(\cdot|k)$ and results from applying the inputs $u^*(\cdot|k)$ to the system starting from the initial state $x(k)$.

The sum of the predicted cost is only minimized over a finite prediction horizon $N \in \mathbb{N}_{>0}$ in (4.2a). This is not only the common case in MPC, but also reasonable under the assumption that the production system changes over time due to unknown or uncertain external influences. Due to such disturbances, the predictions for the distant future are unlikely to be accurate. From the optimal input trajectory $u^*(\cdot|k)$, only the first input $u^*(0|k)$ is applied to the system, before the optimization is repeated with the newly measured state $x(k+1)$ at the next time instant. As introduced in Section 2.4, the resulting nominal closed loop dynamics can be expressed as

$$x(k+1) = Ax(k) + Bu^*(0|k) = x^*(1|k), \qquad x(0) = x^0, \tag{4.3}$$

where a feedback mechanism is introduced through the repeated solution of the optimization problem (4.2). This inherent feedback mechanism is one of the key features of the presented approach compared with many scheduling techniques from literature [80, 91]. It is necessary as the state $x(1|k)$, which was predicted for time $k+1$ starting from $x(k)$ and applying $u^*(0|k)$, might be different from the true state $x(k+1)$. This is caused by external influences and non-modeled effects and has the consequence that the real closed loop might not coincide with the predicted one, i.e., $x(k+1) \neq x(1|k)$.

In order to provide a proof that the proposed MPC scheme leads to a desirable solution of the scheduling problem defined in Section 3.2, we first need to show important properties. Repeatedly applying the optimal first input determined through a finite horizon optimal control problem such as (4.2) does not naturally lead to an optimal closed loop performance [58], which in our case is the completion of all jobs in the production problem in a way that causes the least possible cost. In fact, quite the opposite can be the case, and the MPC scheme might not even start a single task if the MPC problem (4.2) is not formulated carefully and does not fulfill some required conditions, as for example illustrated by Müller and Grüne [57]. In this respect, the MPC problem is analyzed in the following sections and guarantees are provided that the desired properties, i.e., always having a feasible solution and leading to a state in which the production problem is completed, are fulfilled if some specific conditions are satisfied. We first show that there always exists a solution to the optimization problem (4.2) in the following section, before we provide conditions on the cost function $c(x, u)$ and the prediction horizon $N$ which guarantee that applying the control law resulting from the MPC scheme to the manufacturing system will lead to the completion of every task in every job and therefore to the completion of the scheduling problem.

## 4.1.1 Feasibility of the MPC Problem without Terminal Cost

As a first important property of the MPC problem formulated in Section 4.1, it needs to be guaranteed that the optimization problem (4.2) always has a feasible solution when it is applied in closed loop (4.3), which is called *recursive feasibility* in the context of MPC [67]. Proving recursive feasibility of an MPC problem usually requires prior assumptions. In our case, the standing assumption is that the PN was generated with

Algorithms 3 and 4 as described in Section 3.3 for a manufacturing system as described as a JMPMSF as introduced in Section 3.2. We first show that under this assumption the MPC optimization problem (4.2) is feasible for every admissible state of the PN, before specifically analyzing the closed loop system (4.3).

**Lemma 2** (Initial Feasibility). *For every Petri net $PN = (\mathbb{P}, \mathbb{T}, \mathbb{E}, w, x^0)$ generated with Algorithms 3 and 4 represented in its state space form (3.2) the MPC optimization problem (4.2) is feasible for all $x(k) \in \mathbb{N}^n$.*

*Proof of Lemma 2.* The input trajectory $u(\cdot|k) = (0, \ldots, 0)$ satisfies (4.2d), and for all $x(k) \in \mathbb{N}^n$ it satisfies constraint (4.2c) due to Property (P7). Constraint (4.2b) is satisfied with $x(\bar{k}+1|k)$ according to the open loop system dynamics (3.2a). Therefore $u(\cdot|k) = (0, \ldots, 0)$ is a feasible candidate solution of the MPC Problem (4.2) for every $x(k) \in \mathbb{N}^n$. □

For the PN generated with the Algorithms 3 and 4 it particularly holds that $x^0 \in \mathbb{N}^n$ due to Property (P3) and therefore the MPC optimization problem (4.2) is feasible in the initial state $x^0$.

**Lemma 3** (Recursive feasibility). *For every Petri net $PN = (\mathbb{P}, \mathbb{T}, \mathbb{E}, w, x^0)$ generated with Algorithms 3 and 4 represented in its state space form (3.2), the MPC optimization problem (4.2) is feasible at $k = 0$ with $x(0) = x^0$, the nominal closed loop system (4.3) with $u^*(0|k)$ from problem (4.2) satisfies the condition (3.2b), and Problem (4.2) is feasible for all $k \in \mathbb{N}$.*

*Proof of Lemma 3.* From Lemma 2 and as $x^0 \in \mathbb{N}^n$ due to Property (P3) follows that the input trajectory $u(\cdot|k) = (0, \ldots, 0)$ is a feasible candidate solution at $x(0) = x^0$.

That the nominal closed loop system (4.3) with $u(k) = u^*(0|k)$ from problem (4.2) satisfies the condition (3.2b) follows directly from the fact that $x(k) = x(0|k) \in \mathbb{N}^n$ holds in the nominal case and since $u^*(0|k)$ satisfies (4.2c).

For the recursive feasibility, meaning that Problem (4.2) is feasible for all $k \in \mathbb{N}$, we use induction arguments. We show that assuming feasibility at time instant $k$ implies feasibility at instant $k + 1$. Initial feasibility with $x(0|k) = x^0 \in \mathbb{N}^n$ holds according to Lemma 2.

For every feasible input trajectory $u(\cdot|k)$ at time instant $k$, the input trajectory $u(\cdot|k + 1) = (u(1|k), \ldots, u(N - 1|k), 0)$ is a feasible candidate input trajectory to Problem (4.2) at time $k + 1$ with the initial state $x(0|k + 1) = x(k + 1) = x(1|k)$ from the nominal closed loop system (4.3). The feasibility of the inputs $u(0|k + 1) = u(1|k), \ldots, u(N - 2|k + 1) = u(N - 1|k)$ follows directly from the feasibility of $u(\cdot|k)$ at time instant $k$.

The final predicted input $u(N - 1|k + 1) = 0$ is feasible due to the following reasons. First, it naturally satisfies constraint (4.2d). Second, the predicted state $x(N - 1|k + 1)$ with the predicted input $u(N - 1|k + 1) = 0$ satisfies the constraint (4.2b) according to the open loop system dynamics (3.2a). Third, the predicted state $x(N - 1|k + 1)$ is a vector of natural numbers, i.e., $x(N - 1|k + 1) \in \mathbb{N}^n$, due to the feasibility of $u(N-2|k+1) = u(N-1|k)$. For $x(N-1|k+1) \in \mathbb{N}^n$, constraint (4.2c) is satisfied with $u(N-1|k+1) = 0$ due to Property (P7).    □

The input trajectory $u(\cdot|k) = (0, \ldots, 0)$ used in the proof of Lemma 2 simply does not start any task in the scheduling problem. This is surely not desired but in the problem described in Chapter 3.2 there is no external driver prohibiting this solution. In the same sense, extending an existing feasible input trajectory $u(\cdot|k)$ by a new predicted input $u(N-1|k+1) = 0$ in the candidate trajectory for the next time step used in the proof of Lemma 3 does not plan to start any new task in the last step of the prediction.

As explained in Section 3.2.3, there is a production related objective that requires taking some action due to its economic motivation. This economic objective needs to be formulated in the form of a cost function $c(x, u)$ depending on the state $x$ and the input $u$. This leads to the fact that the candidate solutions in the Lemmas 2 and 3 will not be the optimal solutions to the MPC problem (4.2). However, as the optimal open-loop input trajectory $u^*(\cdot|k)$ for the finite horizon optimal control problem (4.2) is not completely applied to the system but only the optimal first input $u^*(0|k)$, the resulting closed loop input trajectory might still not apply any non-zero input.

In the following, we will investigate properties of such a cost function, which we assume to be known and given, that allow to adjust the prediction horizon $N$ such that the production problem controlled with the optimal control law resulting from the MPC scheme is guaranteed to be completed in closed loop. This means that we will guarantee that the closed loop the system will always enter a state $x(k)$ in which the first input $u^*(0|k)$ in the optimal input trajectory $u^*(\cdot|k)$ to the MPC problem (4.2) is not zero, i.e., $u^*(0|k) \neq 0$. In this process we exploit the hierarchical nature of the production problem that is composed of jobs, which themselves are composed of tasks. We first provide a sufficient condition for the completion of a single operation, i.e., a single task of a single job. The arguments in the proof of the completion of an operation motivate the formulation of a more efficient MPC scheme in Section 4.2. After that, we provide a condition for the completion of a job and finally of the whole production problem for both MPC schemes in Section 4.3.

## 4.1.2 Completion of an Operation by the MPC without Terminal Cost

Our goal is to achieve a good performance of the system in closed loop through applying the optimal control law resulting from the MPC scheme. For a flexible manufacturing system (FMS), the first and most important goal is to produce the requested products. In the formulation of the production problem as a JMPMSF in Section 3.2, this means the completion of all the tasks in all jobs. In the model generated in Section 3.3, this is represented as the convergence to a state in which all completion places are marked. A second goal in an FMS is to minimize production cost or to maximize profit. This can be conveniently formulated with a cost function. In the MPC, this cost function is directly use as objective function of the optimization problem (4.2a). As the cost function is only optimized over a limited prediction horizon $N$, the reward of performing actions in the manufacturing system needs to be perceived through evaluating the cost function during this limited time frame. This results in a condition on the cost function $c(x, u)$ and subsequently in a condition on the prediction horizon $N$. In the following, we will show how the prediction horizon $N$ must be chosen based on the given manufacturing system as described in Section 3.2 and a given cost function $c(x, u)$ in order to guarantee completion of a specific operation $O = (\tau, J)$ in the production problem with the MPC formulation from Section 4.1. To achieve this result, we first give a general condition that must hold such that the MPC completes *any* operation.

**Lemma 4** (Completion some operation). *Given a Petri net of a flexible manufacturing system in its state space form* (3.2) *generated with Algorithms 3 and 4. If for a steady state $x^s$ all optimal input trajectories $u^*(\cdot|k)$ for the MPC problem* (4.2) *initialized at $x(k) = x^s$ are such that $u^*(0|k) \neq 0$, then at least one operation $O = (\tau, J)$ will be started at time instant $k$ in closed loop* (4.3), *this operation $O$ will eventually be completed, and it will remain completed.*

*Proof of Lemma 4.* In Algorithm 4, the only transitions that are added to the set of controlled transitions $\mathbb{T}_C$ are starting transitions $T_{(...,S)}$. Therefore, every optimal first input $u^*(0|k) \neq 0$ fires at least one starting transition and thereby initiates the execution of at least one operation by marking at least one production place $P_{(...,P_1)}$. Since we assume $u^*(0|k) \neq 0$ for *all* optimal first inputs, some starting transition is planned to be fired at time instant $k$.

The first input $u^*(0|k)$ is applied to the closed loop system (4.3) controlled by the MPC scheme (4.2) at time instant $k$. Therefore the production is not only started in the prediction, but also in closed loop.

According to Property (P2), the resulting state is no steady state as a production place $P_{(...,P_1)}$ is marked. Due to Property (P8), eventually another steady state is reached. According to Property (P2), a steady state is only reached when no production place $P_{(...,P)}$ is marked any more. As the only independent transitions $T \in \mathbb{T}_I$ that do not mark another production place are finishing transitions $T_{(...,F)}$, this means that eventually a finishing transition fires. As every finishing transition $T_{(...,F)}$ created in Lines 10 and 11 of Algorithm 4 marks a completion place $P_{(...,C)}$, eventually a completion place is marked. Due to Property (P6), once completion places are marked they stay marked and hence the started operation is completed and stays completed. □

The condition in Lemma 4 will be exploited to guarantee the completion of a given production problem as long as it is finite and feasible, which we state in the following assumptions. As we generally allow the production problem to change, for example, through new jobs in the production problem, we assume that it is finite and feasible at the time when we investigate it. For a production problem that can always be extended, trying to argue that it will be completed a certain point in time despite changes is futile. Therefore, we will only analyze the properties of the production problems as it is at a certain point in time.

**Assumption 1** (Finite production system). *The given production problem formulated as in Section 3.2 is finite at a given time, meaning that it has a finite number of possible tasks $n_\mathrm{T} < \infty$, a finite number of jobs $n_J < \infty$ and a finite number of machines $n_\mathrm{M} < \infty$.*

In order to be able to proof completion of a given production problem, we state another central assumption saying that the production problem is feasible meaning that all operations can be executed by the given machine pool. This assumption is formulated on the level of the initial production problem introduced in Section 3.2 as well as on the level of the Petri net and its state space description described in Section 3.3.

**Assumption 2** (Feasibility of the production problem [86]). *All jobs $J \in \mathcal{J}$ can be fulfilled by the given production system. This means that for every job $J \in \mathcal{J}$ and every task $\tau \in \mathcal{T}_J$ there exists at least one machine $M \in \mathcal{M}$ with $\tau \in \mathcal{T}_M$, and that for every task $\tau \in \mathcal{T}_J$ of all jobs $J \in \mathcal{J}$ it holds $\mathcal{T}_\tau \subset \mathcal{T}_J$.*

*In terms of the Petri net, there exists at least one state $\hat{x}$ in which every completion place $P_{(\ldots,C)}$ is marked and which is reachable from the initial state $x^0$.*

*In the state space, there exists at least one state $\hat{x}$ with $\hat{x}_i = 1$ for every $i$ where $x_i^{\mathrm{ID}} = (\ldots, C)$ and which is reachable from the initial state $x^0$, i.e., there exists a feasible input trajectory $(u(0), \ldots, u(\hat{k}-1))$ leading from $x(0) = x^0$ to $\hat{x} = x(\hat{k}) = A^{\hat{k}} x^0 + \sum_{l=0}^{\hat{k}-1} A^l B u(\hat{k}-1-l)$ and where the non-negativity constraint (3.2b) is respected for all $x(k), u(k)$ with $k = 0, \ldots, \hat{k} - 1$.*

**Lemma 5** (Completion of the production problem). *Given a Petri net of a flexible manufacturing system in its state space form (3.2) generated with Algorithms 3 and 4 that fulfills Assumptions 1 and 2. If there is only one steady state $x^\mathrm{F}$ for which not all optimal input trajectories $u^*(\cdot|k)$ for the MPC problem (4.2) are such that $u^*(0|k) \neq 0$, then the production problem will be completed by the MPC scheme and $x^\mathrm{F}$ is the final state of production system.*

*Proof of Lemma 5.* The steady state $x^\mathrm{F}$ is the only steady state for which the preconditions of Lemma 4 are not fulfilled. For all other steady states $x(0|k) = x^\mathrm{s}$, all optimal input trajectories $u^*(\cdot|k)$ for the MPC problem (4.2) are such that $u^*(0|k) \neq 0$. Therefore, Lemma 4 states that in all steady states except $x^\mathrm{F}$ at least one operation $O$ is started. As the production problem is feasible due to Assumption 2, all operations can be executed and since it is finite due to Assumption 1, there is only a finite number of possible operations to be started, so eventually all operations will have been started.

Due to Property (P8) another steady state will be reached after the start of every operation. As $x^{\mathrm{F}}$ is the only steady state for which the preconditions of Lemma 4 are not fulfilled, it is the only steady state in which the system can remain. □

The condition in Lemma 5 that there only exists a single final state $x^{\mathrm{F}}$ is more restrictive than necessary. In general, also for a fixed production problem there is a set $\mathbb{X}_{\mathrm{f}}$ of final states, in which no further production is possible and all jobs are completed. The convergence to this set can be shown with similar arguments.

The condition in Lemma 5 that there is only one steady state in which the optimal feasible behavior is to stay, implies that there cannot be a deadlock in which no further production steps are possible but the production problem is not yet completed. This is the case since in a deadlock state $x$ there would not be any feasible input $u$ but $u = 0$. If another input $u \neq 0$ would be possible in $x$, another production step would be possible and thus $x$ would not be a deadlock state.

Having stated general conditions when an MPC executes operations and ultimately completes the production problem, we will now formulate more specific conditions on the prediction horizon $N$ and the cost function $c(x, u)$ which guarantee the completion of a particular operation $O = (\tau, J)$. As proposed in [86] and [88], we will use the cost function $c(x, u)$ in the MPC problem (4.2) to determine a sufficiently long prediction horizon $N_O$ to guarantee the completion of the operation $O$ for all prediction horizons $N \geq N_O$. In order to proof the completion of every operation, we first define the sets $\mathbb{X}_{O,\mathrm{S}}$ and $\mathbb{X}_{O,\mathrm{C}}$ to classify the steady states in which an operation $O$ can be started and the steady states in which it was just finished, respectively.

**Definition 3** (Set of starting steady states). *For every $O = (\tau, J)$ with $\tau \in \mathcal{T}_J, J \in \mathcal{J}$, the set $\mathbb{X}_{O,\mathrm{S}}$ is defined as the set of steady states in which the operation $O$ can be directly started and which is reachable from the initial state of the production system $x^0$ through a feasible input trajectory.*

**Definition 4** (Set of completion steady states). *For every $O = (\tau, J)$ with $\tau \in \mathcal{T}_J$, $J \in \mathcal{J}$, the set $\mathbb{X}_{O,\mathrm{C}}$ is defined as the set of steady states right after the operation $O$ was completed starting from a state $x^{O,\mathrm{S}} \in \mathbb{X}_{O,\mathrm{S}}$ without executing any further task.*

In order to transfer the production system from a state $x^{O,\mathrm{S}} \in \mathbb{X}_{O,\mathrm{S}}$ to a state $x^{O,\mathrm{C}} \in \mathbb{X}_{O,\mathrm{C}}$, one transition of a special type has to be fired. With the following two definitions, we specify this type of transitions by a set of starting transitions corresponding to operation $O$ and the set of input vectors firing exactly one such transition.

**Definition 5** (Set of starting transitions of an operation). *For a given operation $O = (\tau, M)$, the set of transitions $\mathbb{T}_{O,\mathrm{S}}$ starting the operation $O = (\tau, M)$ is defined as*

$$\mathbb{T}_{O,\mathrm{S}} := \left\{ T_{(M',M,\tau',\tau,J,\mathrm{S})} \in \mathbb{T} \mid M, M' \in \mathcal{M}, \tau' \in \mathcal{T}_{M'} \cap \mathcal{T}_J, \tau \in \mathcal{T}_M \cap \mathcal{T}_J, J \in \mathcal{J} \right\}.$$

**Definition 6** (Set of input vectors starting an operation). *For a given operation $O = (\tau, M)$, the set $\mathbb{U}_{O,\mathrm{S}}$ of inputs starting only the execution of operation $O$ is defined as the set of unit vectors with all entries being zero except for one entry $u_i = 1$ at the position $i$ where $u_i^{\mathrm{ID}} = (M', M, \tau', \tau, J, \mathrm{S})$ and $T_{(M', M, \tau', \tau, J, \mathrm{S})} \in \mathbb{T}_{O,\mathrm{S}}$.*

With those definitions, we state the following theorem to guarantee the completion of an investigated operation $O$ if a condition on the cost function $c(x, u)$ with respect to the starting states $x^{O,\mathrm{S}} \in \mathbb{X}_{O,\mathrm{S}}$ and the resulting completion states $x^{O,\mathrm{C}} \in \mathbb{X}_{O,\mathrm{C}}$ is fulfilled.

**Theorem 1** (Completion of a particular operation $O$ [88]). *Given a Petri net of a flexible manufacturing system in its state space form (3.2) generated with Algorithms 3 and 4 that fulfills Assumptions 1 and 2. If for an operation $O = (\tau, J)$ and a cost function $c(x, u)$ holds that for every state $x^{O,\mathrm{S}} \in \mathbb{X}_{O,\mathrm{S}}$ there exists a state $x^{O,\mathrm{C}} \in \mathbb{X}_{O,\mathrm{C}}$ reachable from $x^{O,\mathrm{S}}$ with*

$$c(x^{O,\mathrm{S}}, 0) > c(x^{O,\mathrm{C}}, 0) \tag{4.4}$$

*then there exists a sufficiently long prediction horizon $N_O \in \mathbb{N}_{>0}$ with the property that for every prediction horizon $N \geq N_O$ the operation $O$ will eventually be completed when starting from any $x^0 \in \mathbb{X}_{O,\mathrm{S}}$ and applying the optimal solution to the MPC problem (4.2) in closed loop (4.3).*

*Proof of Theorem 1.* As shown in Lemma 4, an optimal input trajectory $u^*(\cdot|k)$ with the first predicted input $u^*(0|k) \neq 0$ is required such that some operation is started. In order to prove that the investigated operation $O$ is started, we need to show that the closed loop system always enters a state in which the optimal input trajectory has an input $u^*(0|k) \in \mathbb{U}_{O,\mathrm{S}}$ as first planned input, where the set $\mathbb{U}_{O,\mathrm{S}}$ is defined as in Definitnion 6.

*Part I: Only operation $O$ can be started*
In this part of the proof, we only investigate a point in time when the system is in a state $x$, in which the operation $O$ is the only operation that can be started and show that it is planned to be started immediately in the optimal input trajectory, i.e., $u^*(0|k) \in \mathbb{U}_{O,\mathrm{S}}$, if the prediction horizon is long enough.

We now assume that the considered operation $O = (\tau, J)$ is the *only* operation that can be started and also the only one that needs to be executed in $x^{O,\mathrm{S}}$. For this case, we split all possible predicted trajectories into four categories and compare their resulting cost over the prediction horizon $N$. Without loss of generality, we consider all cases to lead towards the same state $x^{O,\mathrm{C}} \in \mathbb{X}_{O,\mathrm{C}}$, for which condition (4.4) holds with respect to $x^{O,\mathrm{S}}$, except the fourth case where this exception it explicitly mentioned.

1) The operation $O$ is *not executed at all* and thus the first predicted input is $u(0|k) \notin \mathbb{U}_{O,\mathrm{S}}$. As no other operation can be executed, the resulting cost over the prediction horizon $N$ is

$$c_{N,0} = N \, c(x^{O,\mathrm{S}}, 0). \tag{4.5}$$

2) The operation $O$ is planned to *start immediately*, i.e., at $\bar{k} = 0$, on some machine $M \in \mathcal{M}$ with $\tau \in \mathcal{T}_M$ leading to the steady state $x^{O,\mathrm{C}}$. In this case, the first predicted input is $u(0|k) \in \mathbb{U}_{O,\mathrm{S}}$. If no other operation is executed after finishing the execution of the operation $O$, the resulting cost over the prediction horizon $N$ is

$$c_{N,1} = c_\mathrm{P}(M,O) + \big(N - k_\mathrm{P}(M,\tau)\big)\, c(x^{O,\mathrm{C}},0), \qquad (4.6)$$

where $c_\mathrm{P}(M,O)$ is the production cost and $k_\mathrm{P}(M,\tau)$ the production time of the operation $O$ on machine $M$, respectively.

3) The operation $O$ is planned to *start at a later point in time*, assume at $\bar{k} = \hat{k}$, on some machine $M \in \mathcal{M}$ with $\tau \in \mathcal{T}_M$, but it is still completely executed and the steady $x^{O,\mathrm{C}}$ is reached in the prediction. In this case, it holds for the input $u(\hat{k}|k) \in \mathbb{U}_{O,\mathrm{S}}$, but the first predicted input is $u(0|k) \notin \mathbb{U}_{O,\mathrm{S}}$. The resulting cost over the prediction horizon $N$ is

$$c_{N,2} = c_\mathrm{P}(M,O) + \hat{k}\, c(x^{O,\mathrm{S}},0) + \big(N - k_\mathrm{P}(M,\tau) - \hat{k}\big)\, c(x^{O,\mathrm{C}},0) \qquad (4.7)$$

as no other operation can be started in $x^{O,\mathrm{S}}$ and thus the system stays in $x^{O,\mathrm{S}}$ for $\hat{k}$ time steps.

4) The operation $O$ is planned to *start at an even later point in time* on some machine $M' \in \mathcal{M}$ with $\tau \in \mathcal{T}_{M'}$, such that it is not completely executed any more but only the first $\tilde{k}$ production steps, and no steady is reached at the end of the prediction. As no steady state is reached, the execution of $O$ on $M'$ might or might not lead to $x^{O,\mathrm{C}}$ after the prediction horizon. In this case, the input $u(N - \tilde{k}|k) \in \mathbb{U}_{O,\mathrm{S}}$, but the first predicted input is $u(0|k) \notin \mathbb{U}_{O,\mathrm{S}}$. The resulting cost over the prediction horizon $N$ is

$$c_{N,3} = \tilde{c}_\mathrm{P}(M',O) + (N - \tilde{k})\, c(x^{O,\mathrm{S}},0), \qquad (4.8)$$

where $\tilde{c}_\mathrm{P}(M',O)$ is the cost of the first part of the execution of $O$ on $M'$. No other operation can be started in $x^{O,\mathrm{S}}$ and thus the system stays in $x^{O,\mathrm{S}}$ for $N - \tilde{k}$ time steps.

As the MPC scheme (4.2) minimizes the cost over the prediction horizon $N$, the optimal first input $u^*(0,k)$ of the case with the lowest cost will be applied to the system. This means that the operation $O$ will be executed in closed loop if $c_{N,0} > c_{N,1}$, $c_{N,2} > c_{N,1}$ and $c_{N,3} > c_{N,1}$, which means that the immediate execution of the operation $O$ yields the most benefit over the prediction horizon $N$. As the existence of a lower bound $N_O$ for the prediction horizon needs to be proven, we can assume $N \geq N_O > k_\mathrm{P}(M,\tau)$ for all $M$ with $\tau \in \mathcal{T}_M$. With this assumption, also the effect of the production cost $c_\mathrm{P}(M,O)$ and the number of production steps $k_\mathrm{P}(M,O)$ will be negligible, as the following considerations show. First note that due to condition (4.4) and since for the later starting instant $\hat{k} > 0$ it always holds that

$$
\begin{aligned}
c_{N,2} \quad &> \quad c_{N,1} \\
\Leftrightarrow \qquad c_{\mathrm{P}}(M,O) + \hat{k}\, c(x^{O,\mathrm{S}},0) &+ \big(N - k_{\mathrm{P}}(M,\tau) - \hat{k}\big)\, c(x^{O,\mathrm{C}},0) \\
&> \quad c_{\mathrm{P}}(M,O) + \big(N - k_{\mathrm{P}}(M,\tau)\big)\, c(x^{O,\mathrm{C}},0) \\
\Leftrightarrow \qquad 0 \quad &< \quad \hat{k}\big(c(x^{O,\mathrm{S}},0) - c(x^{O,\mathrm{C}},0)\big).
\end{aligned}
\tag{4.9}
$$

For $c_{N,0} > c_{N,1}$ to hold, it follows with Equations (4.5) and (4.6)

$$
\begin{aligned}
N\, c(x^{O,\mathrm{S}},0) \quad &> \quad c_{\mathrm{P}}(M,O) + \big(N - k_{\mathrm{P}}(M,\tau)\big)\, c(x^{O,\mathrm{C}},0) \\
\Leftrightarrow \qquad N \quad &> \quad \frac{c_{\mathrm{P}}(M,O) - k_{\mathrm{P}}(M,\tau)\, c(x^{O,\mathrm{C}},0)}{c(x^{O,\mathrm{S}},0) - c(x^{O,\mathrm{C}},0)}.
\end{aligned}
\tag{4.10}
$$

For $c_{N,3} > c_{N,1}$ to hold, it follows with Equations (4.6) and (4.8)

$$
\begin{aligned}
\tilde{c}_{\mathrm{P}}(M',O) + (N - \tilde{k})\, c(x^{O,\mathrm{S}},0) &> c_{\mathrm{P}}(M,O) + \big(N - k_{\mathrm{P}}(M,\tau)\big)\, c(x^{O,\mathrm{C}},0) \\
\Leftrightarrow \qquad N &> \frac{c_{\mathrm{P}}(M,O) - \tilde{c}_{\mathrm{P}}(M',O) + \tilde{k}\, c(x^{O,\mathrm{S}},0) - k_{\mathrm{P}}(M,\tau) c(x^{O,\mathrm{C}},0)}{c(x^{O,\mathrm{S}},0) - c(x^{O,\mathrm{C}},0)}.
\end{aligned}
\tag{4.11}
$$

As the right hand sides of (4.10) and (4.11) only depend on system parameters that are assumed to be known, these inequalities can always be fulfilled with a large enough value of $N$.

In the cases where a final steady state is reached, all trajectories that lead to a steady state $x^{\mathrm{C},1}$ with higher cost as $x^{O,\mathrm{C}}$, i.e., $c(x^{\mathrm{C},1},0) > c(x^{O,\mathrm{C}},0)$, clearly lead to higher cost over a large prediction horizon $N$ and will not be the optimal input trajectory $u^*(\cdot|k)$.

In the fourth case, choosing a trajectory that does not lead to $x^{O,\mathrm{C}}$ only changes $\tilde{c}_{\mathrm{P}}(M',O)$. From (4.11) it can be seen that an arbitrary, possibly even negative, production cost of the first production steps $\tilde{c}_{\mathrm{P}}(M',O)$ can still be compensated by increasing $N$.

In the cases 2) and 3), we assume that no other operation is planned to start after the execution of the investigated operation $O$. If another operation $O'$ is planned to start right after $O$, we can introduce an artificial operation $\hat{O}$ constituting the execution of $O'$ immediately after $O$ and apply the same arguments as above replacing $O$ by $\hat{O}$. If the other operation $O'$ is only executed after staying in $x^{O,\mathrm{C}}$ for some time, due to condition (4.4) the execution of the investigated operation $O$ is started immediately.

As a last detail in Part I of the proof, we need to show that we do not lose any generality by considering all cases to lead towards the same final state $x^{O,\mathrm{C}} \in \mathbb{X}_{O,\mathrm{C}}$. To show this, we assume the opposite, i.e., that in one of the cases another state $x^{O,\mathrm{C},2} \in \mathbb{X}_{O,\mathrm{C}}$ would be reached. In this case either $c(x^{O,\mathrm{C}},0) < c(x^{O,\mathrm{C},2},0)$ and the arguments above hold as stated, or $c(x^{O,C}, 0) = c(x^{O,C,2}, 0)$ and the arguments above hold independent of whether $x^{O,C}$ or $x^{O,C,2}$ is reached, or $c(x^{O,C}, 0) > c(x^{O,C,2}, 0)$ and we can state the same arguments as above replacing $x^{O,C}$ by $x^{O,C,2}$.

*Part II: Also other operations than the operation $O$ can be started*

Now we investigate a point in time when the system is in a state $x$, in which the operation $O$ is not the only operation that can be started. We will show that in this case the MPC scheme will either start operation $O$ immediately or the MPC scheme will enter another state in which operation $O$ can still be started and the arguments in this proof apply recursively.

Suppose there is at least one other operation $O'$ that can be executed at $x^{O,S}$, i.e., $x^{O,S} \in \mathbb{X}_{O',S}$, then there are two possibilities:

A)  $O$ and $O'$ are executed at the same time,

B)  $O$ and $O'$ are not executed at the same time, either because there is no input vector $u$ starting $O$ and $O'$ and satisfying the non-negativity constraint (3.2b), or because only executing one of them leads to a better solution of the MPC problem (4.2), i.e., an input trajectory $u^*(\cdot|k)$ that leads to lower cost over the prediction horizon $N$.

In the case A), we know that the system will enter another steady state $x^{C,2}$ due to Property (P8). To prove the existence of a sufficiently long prediction horizon $N_O$, we have to assume that $x^{C,2}$ is reached during prediction, which can be achieved by increasing $N$. We neither have any information on the cost of the resulting steady state $x^{C,2}$ after the completion of both operations, nor do we have information on the number of production steps and the production cost of both operations at the same time. As for the considerations in the cases in Part I, in which only the operation $O$ can be started, the effects of the number of production steps and the production cost can be compensated by increasing $N$, as we will see in the remainder of this proof. If the cost of the resulting steady state $x^{C,2}$ is smaller than the cost of the initial state $x^{O,S}$, i.e., $c(x^{C,2}, 0) < c(x^{O,S}, 0)$, the same arguments as in Part I prove that leaving $x^{O,S}$ immediately in order to reach $x^{C,2}$ as soon as possible is the optimal choice and thus with $u^*(0|k)$ either $O$, or $O'$, or both are started. If $O$ or both operations are started, the proof is completed. If another operation $O'$ is started with $u^*(0|k) \in \mathbb{U}_{O',S}$, it is not directly guaranteed that $O$ will be started. However, due to Properties (P8) and (P9), the closed loop system will then eventually enter another steady state $x^{O,S,2} \in \mathbb{X}_{O,S}$. In this state, due to the precondition of Theorem 1, there is another steady state $x^{O,C,2} \in \mathbb{X}_{O,C}$, for which all the arguments in this proof apply. As there are only finitely many operations due to Assumption 1, this recursion can only occur finitely often and eventually the system will enter a steady state $x^{O,S,3} \in \mathbb{X}_{O,S}$ in which due to the precondition of Theorem 1 a steady state $x^{O,C,3} \in \mathbb{X}_{O,C}$ exists and the operation $O$ is executed. This happens at latest if $O$ is the only operation left and the arguments in Part I hold. Note that instead of $O'$ also multiple operations at the same time can be considered in this paragraph on the case A).

In the case B), in which the operations $O$ and $O'$ are executed sequentially, either $O$ is executed first and, due to the same arguments as in Part I, will be executed immediately. Or operation $O'$ is executed first, which can only occur if the execution of $O'$ yields more benefit over the prediction horizon $N$ as the execution of $O$. As argued in Part I, for a sufficiently long prediction horizon $N$ this can only be in combination with a steady state $x^{O',C} \in \mathbb{X}_{O',C}$ with $c(x^{O',C}, 0) < c(x^{O,S}, 0)$. Therefore, the arguments used in Part I prove that operation $O'$ is executed immediately through an input vector $u^*(0|k) \in \mathbb{U}_{O',S}$. Due to the Properties (P8) and (P9), the closed loop system will then eventually enter another steady state $x^{O,S,2} \in \mathbb{X}_{O,S}$ for which, due to the precondition of Theorem 1, there is a steady state $x^{O,C,2} \in \mathbb{X}_{O,C}$ satisfying condition (4.4), for which all the arguments in this proof apply. As argued in the case A), this recursion can only occur finitely often due to Assumption 1, which proves the ultimate execution of the operation $O$.

In the case when (4.2a) does not have a unique minimizer, due to the strict inequality in (4.4), increasing the prediction horizon will lead to uniqueness of the minimizer in the different cases of Part I of the proof. If the different minimumizers correspond to different operations, the finiteness of the problem according to Assumption 1 ensures that the arguments in this proof still hold.

That the operation $O$ will eventually be completed and that it stays completed once its execution had been started follows from Properties (P2), (P6) and (P8) as argued in the proof of Lemma 4. $\qquad\square$

Theorem 1 shows that a cost decrease between starting states and completion states drives the production process. Through the arguments to prove Theorem 1, it can be seen that such a cost decrease combined with a prediction horizon $N$ that is sufficiently long, i.e., $N \geq N_O$, incentivizes an immediate start of operations. Therefore, it implicitly leads to a minimization of the production time, or at least to a short production time under consideration of the given cost function. With Theorem 1, a rigorous condition is provided to examine whether a long enough prediction horizon can lead to the guaranteed completion of the investigated operation. To show that some operation $O$ is guaranteed to be completed, for all its possible starting states $x^{O,S} \in \mathbb{X}_{O,S}$ at least one cheaper completion state $x^{O,C} \in \mathbb{X}_{O,C}$ needs to be found. Since the dimensions of $\mathbb{X}_{O,S}$ and $\mathbb{X}_{O,C}$ can be arbitrarily large, this is computationally expensive for arbitrary cost functions. In Section 4.3, we show that this can be done more efficiently if a linear cost function (4.1) is employed.

The proof of Theorem 1 also illustrates, that a cheap initial state combined with cheap production processes and expensive final states such that condition (4.4) is not fulfilled for any final state can lead to the situation in which the optimal predicted trajectory stays in the initial state for some time before the production process is planned to be initiated. As only the optimal first input $u^*(0|k) = 0$ is applied, under those circumstances the closed loop system (4.3) stays in the initial state and no production is started at all.

# 4.2 MPC Formulation with Terminal Cost

Theorem 1 is based on extending the prediction horizon $N$ such that staying at a final steady state $x^{O,C}$ with low cost $c(x^{O,C}, 0)$ for a long amount of time will incentivize the immediate start of an investigated operation $O$. Due to Property (P8), this leads to the completion of the investigated operation. This insight can be used to formulate a more efficient MPC problem with terminal cost that has similar properties:

$$\underset{u(\cdot|k)}{\text{minimize}} \quad \sum_{\bar{k}=0}^{N-1} c\left(x\left(\bar{k}|k\right), u\left(\bar{k}|k\right)\right) + V_{\mathrm{f}}\left(x(N|k)\right) \tag{4.12a}$$

$$\text{subject to} \quad x\left(\bar{k}+1|k\right) = Ax\left(\bar{k}|k\right) + Bu\left(\bar{k}|k\right), \tag{4.12b}$$

$$0 \leq x\left(\bar{k}|k\right) - B^{-}u\left(\bar{k}|k\right), \tag{4.12c}$$

$$u\left(\bar{k}|k\right) \in \mathbb{N}^{m}, \quad \text{for } \bar{k} = 0, \ldots, N-1, \tag{4.12d}$$

$$x\left(0|k\right) = x\left(k\right). \tag{4.12e}$$

In this formulation, the terminal cost $V_{\mathrm{f}} : \mathbb{N}^{n} \to \mathbb{R}$ is introduced with respect to the previous formulation of the MPC problem (4.2). It can be used in order to capture the cost of resting in a final steady state for an extended amount of time. By assigning

$$V_{\mathrm{f}}\left(x\right) = \sum_{\tilde{k}=0}^{N^{+}} c(A^{\tilde{k}}x, 0), \tag{4.13}$$

the cost of assigning the input $u(\bar{k}|k) = 0$ for $\bar{k} = N, \ldots, N + N^{+}$ to the system is evaluated through $V_{\mathrm{f}}\left(x(N|k)\right)$, where we call $N^{+}$ the *extended prediction horizon*. Note that it holds that $\tilde{k} = \bar{k} - N$ for $\bar{k} \geq N$ when the terminal cost function (4.13) is evaluated at the terminal state, i.e., $V_{\mathrm{f}}\left(x(N|k)\right)$. This strategy to enlarge the prediction horizon by a finite amount $N^{+}$ over which the control input is kept constant is a well-known strategy in MPC literature [1]. It is known that capturing the cost over this extended period of time in a terminal cost can be used to prove stability of the closed loop system. Due to Property (P8), for a sufficiently long extended prediction horizon $N^{+}$ the system approaches a steady state and stays there as long as the input $u(\bar{k}|k) = 0$ is applied. With this insight, we can formulate the following corollary.

**Corollary 1** (Completion of an operation with the MPC scheme with terminal cost). *Given a Petri net of a flexible manufacturing system in its state space form (3.2) generated with Algorithms 3 and 4 that fulfills Assumptions 1 and 2. If for an operation $O = (\tau, J)$ and a cost function $c(x, u)$ holds that for every state $x^{O,S} \in \mathbb{X}_{O,S}$, there exists a state $x^{O,C} \in \mathbb{X}_{O,C}$ reachable from $x^{O,S}$ with*

$$c(x^{O,S}, 0) > c(x^{O,C}, 0) \tag{4.14}$$

*then there exists a sufficiently long extended prediction horizon* $N_O^+ \in \mathbb{N}_{>0}$ *with the property that for every* $N^+ \geq N_O^+$ *and every* $N \in \mathbb{N}_{>0}$ *the operation* $O$ *will eventually be executed when starting from any* $x^0 \in \mathbb{X}_{O,S}$ *and applying the optimal solution to the MPC problem* (4.12) *with the terminal cost* (4.13) *in closed loop* (4.3).

*Proof of Corollary 1.* The proof is analogous to the proof of Theorem 1. First, note that staying in the final state $x^{O,C}$ is beneficial compared with staying in $x^{O,S}$ due to condition (4.14). Due to Property (P8), this final state will be reached in the prediction as soon as an input $u(0|k) \in \mathbb{U}_{O,S}$ starting the operation $O$ is applied and $u(\bar{k}|k) = 0$ for $\bar{k} > 0$. The terminal cost $V_f$ defined in (4.13) is the cost of finishing all running production processes and then staying at the final steady state $x^{O,C}$ by applying $u(\bar{k}|k) = 0$ for all $\bar{k} = N, \ldots, N + N^+$. With every prediction horizon $N > 0$, the MPC scheme selects the best possible inputs during the first $N$ steps considering the autonomous continuation of the processes for further $N^+$ time steps after the prediction horizon.

In every prediction horizon $N > 0$ starting at a state $x^{O,S} \in \mathbb{X}_{O,S}$, the possibility of immediately starting a process leading from $x^{O,S}$ to $x^{O,C}$, i.e., applying an input $u^*(0|k) \in \mathbb{U}_{O,S}$, is considered in the optimization. If the horizon $N + N^+$, over which the cost is considered, is long enough, it follows from the same arguments as in the proof of Theorem 1 that applying the input $u^*(0|k) = 0$ and not starting any operation cannot be more beneficial than immediately starting some operation $O'$. Due to condition (4.14), one beneficial possibility would be starting $O' = O$ with $u^*(0|k) \in \mathbb{U}_{O,S}$. As there might be other beneficial inputs $u^*(0|k) \in \mathbb{U}_{O',S}$ starting other operations $O'$, the operation that is immediately started does not have to be the investigated operation $O$. As soon as $N + N^+$ is large enough, however, a condition as (4.14) also needs to hold for such an other operation $O'$ if it is optimal to start it through an input $u^*(0|k) \in \mathbb{U}_{O',S}$. Therefore, as shown in the proof of Theorem 1, it is always better to start such an operation immediately compared with starting it later. Due to Properties (P8) and (P9) and finiteness of the production problem assumed in Assumption 1, it is guaranteed that at some point in time the closed loop (4.3) reaches a state in which applying $u^*(0|k) \in \mathbb{U}_{O,S}$ is the optimal solution of the MPC problem (4.12) and the operation $O$ will be started.

That the operation $O$ will eventually be completed and that it stays completed once its execution had been started follows from Properties (P2), (P6) and (P8) as argued in the proof of Lemma 4. $\qquad\square$

# 4.3 Completion of the Production Problem

In the previous sections, we have proven the completion of an operation $O$ when either the MPC scheme (4.2) or the MPC scheme with the terminal cost (4.12) and the terminal cost function (4.13) is applied in closed loop, which is formulated in Theorem 1 and Corollary 1. This is the basis to guarantee the completion of every job and therefore of the entire production problem.

**Corollary 2** (Completion of a job). *Given a Petri net of a flexible manufacturing system in its state space form* (3.2) *generated with Algorithms 3 and 4 that fulfills Assumptions 1 and 2. If for every task $\tau \in \mathcal{T}_J$ in a job $J$ there exists*

(a) *a sufficiently long prediction horizon $N_O$ for the operation $O = (\tau, J)$, then there exists a sufficiently long prediction horizon $N_J = \max_{\tau \in \mathcal{T}_J} N_O$ with the property that for every $N \geq N_J$ the job $J$ will eventually be completed when applying the optimal control law resulting from the MPC scheme* (4.2) *in closed loop* (4.3) *from any initial state $x^0 \in \mathbb{X}_{J,S} := \bigcup_{\tau \in \mathcal{T}_J} \mathbb{X}_{O,S}$.*

(b) *a sufficiently long extended prediction horizon $N_O^+$ for the operation $O = (\tau, J)$, then there exists a sufficiently long extended prediction horizon $N_J^+ = \max_{\tau \in \mathcal{T}_J} N_O^+$ with the property that for every $N \in \mathbb{N}_{>0}$ and every $N^+ \geq N_J^+$ the job $J$ will eventually be completed when applying the optimal control law resulting from the MPC scheme* (4.12) *with the terminal cost* (4.13) *in closed loop* (4.3) *from any initial state $x^0 \in \mathbb{X}_{J,S} := \bigcup_{\tau \in \mathcal{T}_J} \mathbb{X}_{O,S}$.*

*Proof of Corollary 2.* We only prove the case (a) for the MPC scheme (4.2) in detail. The case (b) for the MPC scheme with extended prediction horizon (4.12) and terminal cost (4.13) follows directly with the same arguments as in the proof of Corollary 1.

From Theorem 1 it follows that, if for an operation $O$ there exists a sufficiently long prediction horizon $N_O$, the operation $O$ will eventually be completed when applying the MPC scheme (4.2) once the system has entered an appropriate state $x^{O,S} \in \mathbb{X}_{O,S}$ as long as the prediction horizon $N$ is long enough, in particular if $N \geq N_O$. Since a sufficiently long prediction horizon $N_O$ is assumed to exist for every operation $O = (\tau, J)$ with $\tau \in \mathcal{T}_J$, and since it is assumed that $N \geq N_J \geq N_O$ for all such operations, this holds for all investigated operations.

The system starts in an initial state $x^0 \in \mathbb{X}_{J,S} = \bigcup_{\tau \in \mathcal{T}_J} \mathbb{X}_{O,S}$, in which at least one operation $O = (\tau, J)$ with $\tau \in \mathcal{T}_J$ can be started directly. Due to Theorem 1 this operation will eventually be completed. Assumption 2 ensures that after the completion of every operation $O$ there is at least one other operation $O'$ that can be executed, i.e., $x^{O,C} \in \mathbb{X}_{O',S}$, except all operations of the investigated job $J$ are already completed. Due to the finite nature of the problem according to Assumption 1, this will eventually happen. □

**Corollary 3** (Completion of the production problem). *Given a Petri net of a flexible manufacturing system in its state space form* (3.2) *generated with Algorithms 3 and 4 that fulfills Assumptions 1 and 2. If for every job $J \in \mathcal{J}$ there exists*

(a) *a sufficiently long prediction horizon $N_J$, then there exists a sufficiently long prediction horizon $N_{\min} = \max_{J \in \mathcal{J}} N_J$ with the property that for every $N \geq N_{\min}$ the given production problem will eventually be completed when applying the optimal control input resulting from the MPC scheme* (4.2) *in closed loop* (4.3) *from any initial state $x^0 \in \mathbb{X}_S := \bigcap_{J \in \mathcal{J}} \mathbb{X}_{J,S}$.*

*(b)* *a sufficiently long extended prediction horizon $N_J^+$, then there exists a sufficiently long extended prediction horizon $N_{\min}^+ = \max_{J \in \mathcal{J}} N_J^+$ with the property that for every $N \in \mathbb{N}_{>0}$ and every $N^+ \geq N_{\min}^+$ the given production problem will eventually be completed when applying the optimal control input resulting from the MPC scheme (4.12) with the terminal cost (4.13) in closed loop (4.3) from any initial state $x^0 \in \mathbb{X}_S := \bigcap_{J \in \mathcal{J}} \mathbb{X}_{J,S}$.*

*Proof of Corollary 3.* We again only prove the case (a) for the MPC scheme (4.2) and the case (b) for the MPC scheme with extended prediction horizon (4.12) and terminal cost (4.13) follows directly.

In Corollary 3 it is assumed that for every job $J \in \mathcal{J}$ there exists a sufficiently long prediction horizon $N_J$. From Corollary 2 we know that in this case every job will eventually be completed if the system is initialized in an appropriate initial state $x^0 \in \mathbb{X}_{J,S}$. This is the case for all jobs $J \in \mathcal{J}$ since the system is assumed to be initialized in the intersection of the required sets $\mathbb{X}_{J,S}$ of all jobs $J \in \mathcal{J}$, i.e., $x^0 \in \mathbb{X}_S := \bigcap_{J \in \mathcal{J}} \mathbb{X}_{J,S}$. $\qquad\square$

With the results provided so far, we show the applicability of MPC to the scheduling of an FMS described as JMPMSF. It is guaranteed with Lemma 2 that the MPC schemes are feasible if they are initialized appropriately with a state $x(k) \in \mathbb{N}^n$. Based on that, Lemma 3 shows that the MPC schemes remain feasible when the resulting optimal control law is applied in the nominal closed loop. In Theorem 1 and Corollaries 1 - 3 conditions are provided which allow to parametrize the MPC schemes in a way that ensures the completion of the production problem. The modular structure of the PN model for the JMPMSF is exploited by first determining a sufficiently long prediction horizon $N_O$ or $N_O^+$ for every operation in a job in order to guarantee their completion, if the MPC scheme (4.2) or the MPC scheme with terminal cost (4.12) and the terminal cost function (4.13) is used, respectively. From them, a sufficiently long prediction horizon $N_J$ or $N_J^+$, respectively, for every job of the production problem are determined. The sufficiently long prediction horizons of all jobs can finally be used to compute a sufficiently long prediction horizon $N_{\min}$ or $N_{\min}^+$, respectively, that guarantee the completion of the production problem. How the MPC scheme (4.12) with the terminal cost (4.13) can be applied for reactive scheduling will be shown in Section 5 with two numerical examples. In these examples, a linear cost function is used which simplifies the search for a sufficiently long prediction horizon as we explain in the following section, in which we discuss further properties of the MPC schemes.

## 4.4 Discussion

In the previous sections, we developed two MPC schemes for the reactive scheduling of FMS and proved that they are recursively feasible and will ultimately complete the scheduling problem in closed loop. In this section, we discuss further properties of the MPC schemes in order to illustrate their usefulness for the scheduling of FMSs. In particular, in Section 4.4.1 we stress how the MPC schemes embedded in the framework described in Section 3.1 are able to react to changes in the manufacturing system and how the cost function employed in the MPC problems relates to the original problem description. Section 4.4.2 covers the relation between the cost in the initial problem formulation and the cost function $c(x, u)$ used in the MPC schemes and in Section 4.4.3 we illustrate how providing the guaranteed completion of the production system is simplified if a linear cost function (4.1) is used. In Section 4.4.4, we discuss how alternative techniques form literature can be used to achieve similar convergence properties.

## 4.4.1 Robustness to Changes in the System

For the scheduling problem represented as state space description of a PN, two MPC optimization problems were formulated. The scheduling objective to achieve the best possible performance is directly considered by minimizing the cost that accumulates over time with respect to a given cost function. The optimization is done with respect to a constant system model at first. However, under the assumption that changes are frequent in an FMS and thus the optimization for the far future might be futile, two actions are taken. On the one hand, the optimization is only conducted over a limited time horizon, and on the other hand, the optimization result is not completely applied to the system, but only the actions planned for the current time instant. In order to determine the input, i.e., the manufacturing decisions, for the following time step, the current system state is measured and the optimization is conducted based on the true system state. The measurements are assumed to be readily available through the digital twins of the machines and products. By this, a feedback mechanism is introduced that naturally makes the scheduling strategy more robust and flexible than simply trying to apply an optimal open loop strategy. This general approach of MPC seems particularly suited under the assumption of frequent changes of FMSs in Industry 4.0. In the following, we will highlight some relevant changes in the system and discuss the reaction of the system in the presented framework.

**Disturbances**

If the production problem is given and does not change over time, the MPC formulations can be parametrized such that the completion of the production problem is guaranteed by employing the results in Theorem 1 and Corollaries 1–3. Those guarantees are given with respect to the model of the system. However, due to the way they are determined, for the case of differences between the model and the real system, the controller will at least apply an input to the system and attempt to complete the scheduling problem. Such differences can be represented in the model with a disturbance vector $d(k) \in \mathbb{R}^n$ through

$$x(k + 1) = Ax(k) + Bu(k) + d(k). \tag{4.15}$$

As long as differences between the real system and its model are perceived and the state of the model is updated accordingly through the digital twins, they are considered through the state measurement in the feedback mechanism depicted in Figure 3.1. If the disturbances do not lead to the violation of the required assumptions and conditions of Theorem 1 and Corollaries 1–3, the guarantees are retained. Disturbances of this kind could be the breakdown of a machine, which is fed back by removing a token from the respective production or idle place corresponding to the broken machine in the PN. The removal of a token from a place sets an non-zero entry in the state vector $x$ to zero and the state remains a vector of natural numbers, i.e., $x \in \mathbb{N}^n$. Therefore, according to Lemma 2, the MPC problem remains feasible. As long as at least one of the other machines in the manufacturing system, which is still functional, can take over all the tasks of the broken machine, Assumption 2 is still satisfied. If the preconditions of Theorem 1 are still fulfilled for all operations that have to be taken over by another machine, the convergence guarantees can be retained or at least restored by adapting the prediction horizon $N$ in the case of the MPC formulation (4.2) or adapting $N^+$ in the case of the MPC scheme (4.12). To determine whether an adaptation of $N$ or $N^+$ is necessary during runtime would require additional computational effort. Therefore, in order to save the additional computations during runtime, a prediction horizon which is longer as the shortest sufficient prediction horizon $N_{\min}$ or $N_{\min}^+$ can be chosen. This measure leads to a certain level of redundancy of the possible machines if multiple machines are able to generate profit through executing the same operation. This redundancy makes the closed loop system robust with respect to machine breakdowns. After a defective machine is available again, the previously removed token is restored in the PN in the place where it was removed at the time of the breakdown, and the production is continued as intended.

**Changes in the System**

Due to its modular structure, the model offers the possibility to include new jobs or new machines, as required in Section 3.2.3 and motivated by real production scenarios. This modularity is maintained through the model generation to the formulation of the control problem, and therefore it is easily possible to analyze whether the required properties are preserved despite changes in the problem description. Including a new machine or job results in new system matrices

$$A_{\text{new}} = \left[ \begin{array}{c|c} A_{\text{old}} & A_{\text{old,add}} \\ \hline A_{\text{add,old}} & A_{\text{add,add}} \end{array} \right] \quad \text{and} \quad B_{\text{new}} = \left[ \begin{array}{c|c} B_{\text{old}} & B_{\text{old,add}} \\ \hline B_{\text{add,old}} & B_{\text{add,add}} \end{array} \right] \quad (4.16)$$

and new state and input vectors

$$x_{\text{new}} = \left[ \begin{array}{c} x_{\text{old}} \\ x_{\text{add}} \end{array} \right] \quad \text{and} \quad u_{\text{new}} = \left[ \begin{array}{c} u_{\text{old}} \\ u_{\text{add}} \end{array} \right] \quad (4.17)$$

based on the old matrices and vectors $A_{\text{old}}$, $B_{\text{old}}$, $x_{\text{old}}$ and $u_{\text{old}}$, where the elements $A_{\text{old,add}}$, $A_{\text{add,old}}$, $A_{\text{add,add}}$, $B_{\text{old,add}}$, $B_{\text{add,old}}$, $B_{\text{add,add}}$, $x_{\text{add}}$ and $u_{\text{add}}$ relate to the newly added machine or job and connect it to the existing system. The new matrices for $B^+$ and $B^-$ have an analogous form and the assignment between the existing and the new elements can be realized by means of the vectors of identifiers $x^{\text{ID}}$ and $u^{\text{ID}}$.

The introduction of a new machine changes the convergence guarantees of the MPC schemes given with Theorem 1 and Corollaries 1–3 only in exceptional cases. Generally, a new machine introduces further flexibility and thereby additional possibilities to perform manufacturing tasks in an efficient way without restricting existing ones. To be precise, it is unlikely that a new machine influences the production cost $c_P(\tau, M)$ of a task $\tau$ on an existing machine $M$, or the relation between the cost of existing steady states $x^{O,S}$ and $x^{O,C}$. If it is assumed that the properties of the existing parts of the production problem do not change, the introduction of a new machine does not change the convergence properties of the MPC schemes.

For the introduction of a new job $\hat{J}$, analogous arguments hold and the modular structure of the production problem can be exploited by both, the MPC scheme without terminal cost (4.2) and the MPC scheme with terminal cost (4.12) and the terminal cost function (4.13). The completion of the production problem can be guaranteed, by only analyzing the new job $\hat{J}$. First, for every task $\tau \in \mathcal{T}_{\hat{j}}$ of the new job it is investigated whether a sufficiently long prediction horizon $N_O$ or $N_O^+$, respectively, of the operation $O = (\tau, \hat{J})$ can be found. If this is the case for all operations, a sufficiently long prediction horizon $N_{\hat{j}}$ or $N_{\hat{j}}^+$, respectively, of the new job can be computed according to Corollary 2. If it is shorter than the existing value of the sufficiently long prediction horizon $N_{\min}$ or $N_{\min}^+$, respectively, the convergence properties are preserved despite the new job. If $N_{\hat{j}}$ or $N_{\hat{j}}^+$, respectively, is longer, the sufficiently long prediction horizon can be adjusted accordingly. If necessary, the prediction horizon $N$ or the extended prediction horizon $N^+$ needs to be extended to restore the convergence guarantee despite the new job.

Changes in the cost function, which might be caused by decisions made on the management level or changing energy or material cost, can be handled as long as condition (4.4) in Theorem 1 or condition (4.14) in Corollary 1 are not violated. However, in order to retain the convergence guarantees of the proposed MPC schemes, the evaluation of the used (extended) prediction horizon $N$ ($N^+$) with respect to Theorem 1 (Corollary 1) needs to be repeated in this case.

## 4.4.2 Production Cost Allocation

We formulated the maximization of profitability of the manufacturing system as objective for the presented scheduling scheme in Section 3.2.3. We assumed that it is possible to quantify the profitability with a cost function that can be used in the optimization problems (4.2) and (4.12). In practice, the formulation of a suitable cost function that represents the desired production goals is not an easy task. Even formulating scheduling objectives that are frequently used in literature, as, for example, the optimality criteria presented by Brucker [10], in the form of a stage cost function $c(x, u)$ is not always trivial. On the other hand, it is also not clear whether and under which circumstances the most common scheduling objectives lead to the highest possible profit, as already Manne pointed out [55].

In the previous sections, first the completion of the single operations, then the completion of the jobs and ultimately of the complete production problem was proven if the cost function $c(x, u)$ fulfills condition (4.4) or (4.14) for at least one completion state $x^{O,C} \in \mathbb{X}_{O,C}$ corresponding to every starting state $x^{O,S} \in \mathbb{X}_{O,S}$. This property of the cost function is formulated on the level of the state space description of the PN. Since the original scheduling problem for an FMS designed according to the principles of skill-based engineering is formulated as a JMPMSF as introduced in Chapter 3, this property can only be checked after transforming the problem into a PN. Although the transformation process is done automatically with Algorithms 3 and 4, we did not yet specify how the cost function for the state space description originates from the JMPMSF.

As explained in Section 3.2.3, the cost function should respect the economic nature of the problem and therefore we do not want to completely specify it in this section. Nevertheless, in order to guarantee the completion of the production problem with the introduced MPC schemes (4.2) and (4.12), the cost function has to satisfy condition (4.4) or (4.14), respectively. To provide the required guarantees while striving for economic benefit, condition (4.4) or (4.14) needs to be fulfilled while preserving the economic nature of the original problem. Therefore, when assigning the cost to the elements of the PN, the economic objective and condition (4.4) or (4.14) to guarantee convergence must be considered concurrently. We now discuss to which elements of the PN the economic cost can be assigned and which parts are suited to provide the convergence guarantee.

**Assigning Cost to the Petri Net Elements**

First note that the cost function $c(x, u)$ depends on the state of the PN $x$ corresponding to the markings of the places and on the firing count vector $u$ corresponding to the firing of the controlled transitions. For guaranteeing the completion of an operation $O$ in Theorem 1, the cost of the starting states $x^{O,S} \in \mathbb{X}_{O,S}$ and corresponding completion states $x^{O,C} \in \mathbb{X}_{O,C}$ are considered. Due to their description in the Definitions 3 and 4, they are steady states. The execution of an operation $O = (\tau', J)$ in the PN created with Algorithms 3 and 4, which is initiated through the firing of a starting transition $T_{(M,M',\tau,\tau',J,\mathrm{S})}$ from a state $x^{O,S} \in \mathbb{X}_{O,S}$, depends on places of four different types to be marked. As illustrated in Figure 3.5, those are one idle place $P_{(M',0,0,\mathrm{I})}$, one buffer place $P_{(M,\tau,J,\mathrm{B})}$, one necessity place $P_{(0,\tau',J,\mathrm{N})}$ and possibly multiple completion places $P_{(0,\tau'',J,\mathrm{C})}$, $\tau'' \in \mathcal{T}_{\tau'}$. After the completion of the operation $O$, in the sate $x^{O,C} \in \mathbb{X}_{O,C}$ the idle place $P_{(M',0,0,\mathrm{I})}$ and the completion places $P_{(0,\tau'',J,\mathrm{C})}$, $\tau'' \in \mathcal{T}_{\tau'}$ are marked again. Additionally another buffer place $P_{(M',\tau',J,\mathrm{B})}$ and the completion place $P_{(0,\tau',J,\mathrm{C})}$ are marked. In particular it can be seen that before the start of an operation $O = (\tau', J)$ the necessity place $P_{(0,\tau',J,\text{N})}$ is marked and after its completion the completion place $P_{(0,\tau',J,\text{C})}$ is marked. Both of those places are characteristic for the operation $O = (\tau', J)$. Since this process cannot be reversed as stated in Property (P6), the place $P_{(0,\tau',J,\text{C})}$ stays marked after the production process and $P_{(0,\tau',J,\text{N})}$ will not be marked again. Therefore, a guaranteed cost decrease between the starting state $x^{O,\text{S}}$ and the completion state $x^{O,\text{C}}$ can be realized though the dependence of the cost function $c(x, u)$ on those two places. Assume the two states $x^{O,\text{S}}$ and $x^{O,\text{C}}$ are identical except for the entries corresponding to the places $P_{(0,\tau',J,\text{N})}$, which is marked in $x^{O,\text{S}}$ and unmarked in $x^{O,\text{C}}$, and $P_{(0,\tau',J,\text{C})}$, which is marked in $x^{O,\text{C}}$ and unmarked in $x^{O,\text{S}}$. Then only the cost of the markings of $P_{(0,\tau',J,\text{N})}$ and $P_{(0,\tau',J,\text{C})}$ decide whether the operation $O$ can be guaranteed to be completed through increasing the prediction horizon $N$ according to Theorem 1 starting from $x^{O,\text{S}}$. In this case, the operation $O = (\tau', J)$ can be guaranteed to be completed by artificially adjusting the weight of the places $P_{(0,\tau',J,\text{N})}$ and $P_{(0,\tau',J,\text{C})}$ in the cost function $c(x, u)$ and sufficiently increasing the prediction horizon $N$.

Besides the two places discussed in the previous paragraph, also the pair of buffer places $P_{(M,\tau,J,\text{B})}$ and $P_{(M',\tau',J,\text{B})}$ seems to be a candidate for forcing a cost decrease, as Figure 3.5 illustrates. This, however, is not a valid option in general, since for two tasks $\tau$ and $\tau'$ which have no precedence relation with one another, it is not predetermined which of them will be executed first. Therefore, it is not possible to decide whether $P_{(M,\tau,J,\text{B})}$ or $P_{(M',\tau',J,\text{B})}$ should be assigned the larger cost for being marked. Through artificially increasing the cost of one of those two places over the other, a desired precedence can be introduced. This can be exploited if one production sequence if preferable over the other. If, for example, executing the task $\tau$ before $\tau'$ is preferable from an application point of view, a larger cost for the marking of the place $P_{(M,\tau,J,\text{B})}$ compared with the cost for the marking of $P_{(M',\tau',J,\text{B})}$ will consider this preference in the optimization problems (4.2) and (4.12). However, if other cost terms that are not considered in this reasoning still make executing $\tau$ after $\tau'$ favorable, the MPC schemes might lead to such a solution, which is the economically better choice.

One-time cost in the production process that do not accumulate over time are best represented through the influence of the input vector $u$ in the cost function $c(x, u)$, or through influence of the markings of the production places $P_{(M,\tau,J,\text{P}_q)}$. The firing of a particular controlled transition through a non-zero entry in the input vector $u$ can only happen once during the production process. Similarly, the production places $P_{(M,\tau,J,\text{P}_q)}$ are only marked for a single time instant, before being emptied through the autonomous firing of the production transition $T_{(M,M,\tau,J,\text{P}_q)}$. In the cost related the starting transitions and thereby to the firing count vector $u$, one-time cost associated with the execution of the respective operation can be considered. This can, for example, be the cost of the material used in the initiated task. The cost associated with the production places can for example account for power consumption, which might vary over the production process leading to different cost of the places $P_{(M,\tau,J,\text{P}_q)}$ and $P_{(M,\tau,J,\text{P}_{q+1})}$. Influences as the energy price might depend nonlinearly on the power consumption of the entire production plant [48]. Therefore, isolated considerations of the cost related to the single elements of the PN are only possible in exceptional cases as for example when comparing two states that only differ in the considered PN element, or when the cost function exhibits certain properties such as linearity.

### 4.4.3 Linear Cost Function

The most challenging part in providing the convergence guarantees is to determine a sufficiently long prediction horizon $N_O$ or $N_O^+$ for an operation $O$. For the MPC scheme without terminal cost (4.2), determining $N_O$ requires to find a state $x^{O,C} \in \mathbb{X}_{O,C}$ for all states $x^{O,S} \in \mathbb{X}_{O,S}$ that satisfies condition (4.4). Analogously, for the MPC scheme with terminal cost (4.12), determining $N_O^+$ requires to find a state $x^{O,C} \in \mathbb{X}_{O,C}$ for all states $x^{O,S} \in \mathbb{X}_{O,S}$ that satisfies condition (4.14). Only if such a pair $(x^{O,S}, x^{O,C})$ is found, a value for $N_O$ or $N_O^+$ can be computed. Since for the MPC scheme without terminal cost (4.2) and for the MPC scheme with terminal cost (4.12) the same criteria need to be investigated, we only discuss the MPC scheme without terminal cost (4.2) in this section and the results hold for scheme with terminal cost (4.12) as well.

For arbitrary cost functions, finding a sufficiently long prediction horizon $N_O$ might require to investigate all inputs $u_T \in \mathbb{U}_{O,S}$ that are feasible in one of the states in the set $\mathbb{X}_{O,S}$, where $u_T$ fires a single transition starting the operation $O$ as defined in Definition 6. By exploiting Properties (P4), (P5), and (P8), the states $x^{O,C}$ can be characterized based on the state $x^{O,S}$, an input $u_T \in \mathbb{U}_{O,S}$ and the matrices of the algebraic description of the PN as $x^{O,C} = x^{O,S} - B^- u_T + A^n B^+ u_T$. With this characterization, the condition (4.4) can be expressed without the need to explicitly determine the set $\mathbb{X}_{O,C}$

$$c(x^{O,S}, 0) > c(x^{O,S} - B^- u_T + A^n B^+ u_T, 0). \tag{4.18}$$

This expression results from the fact that the firing of a transition $u_T$ removes tokens form its input places according to $B^- u_T$ and adds tokens to its output place according to $B^+ u_T$. The output $B^+ u_T$ of a starting transition $T$ fired through $u_T \in \mathbb{U}_{O,S}$ is a production place and thus its token is moved forward through the independent part of the PN during the production process. This is captured through the repeated multiplication with the matrix $A$. Finally, after the production process ended, the token created through the firing of $T$ will result in the markings $A^n B^+ u_T$ as stated in Property (P8).

Due to the large cardinality of the set $\mathbb{X}_{O,S}$, it might be practically impossible to check condition (4.18) for all $x^{O,S} \in \mathbb{X}_{O,S}$ to prove the completion of a single operation. In this endeavor, many states $x^{O,S}$ are investigated, which the system will not even attain. Since condition (4.4), and hence also condition (4.18), is only a sufficient condition for the completion of an operation $O$, the effort for providing the theoretical guarantees might be bigger as required.

To facilitate this endeavor, we provide a condition for the case of a linear cost function (4.1), i.e., $c(x, u) = c_x^\top x + c_u^\top u$. This is done by exploiting Property (P4) and the linearity of the cost function, and thereby reducing the number of conditions to be investigated. For a given transition $T$, there are many possible steady states $x^{O,S} \in \mathbb{X}_{O,S}$ in which an input $u_T \in \mathbb{U}_{O,S}$ firing only the transition $T$ can be applied without violating the non-negativity constraint (2.3). For a linear cost function, instead of investigating all such steady states $x^{O,S} \in \mathbb{X}_{O,S}$, only a single condition needs to be checked for each transition $T$ independently of any steady state $x^{O,S}$. Only the precondition of the input $u_T \in \mathbb{U}_{O,S}$ based on the matrix $B^-$ needs to be checked and thereby the number of conditions to investigate is significantly reduced.

Every state $x^{O,S} \in \mathbb{X}_{O,S}$ can be expressed as $x^{O,S} = \bar{x} + B^- u_T$ for some $u_T \in \mathbb{U}_{O,S}$. Since the set $\mathbb{X}_{O,S}$ is defined in Definition 3 such that at least one $u_T \in \mathbb{U}_{O,S}$ is enabled, the remaining vector $\bar{x}$ after firing the transition $T$ through $u_T$ is non-negative, i.e., $\bar{x} \geq 0$. Due to Property (P4), the vector $\bar{x}$, which the steady states $x^{O,S}$ and $x^{O,C}$ have in common, is itself a steady state and remains unchanged during the execution of the operation $O$. The relevant part for enabling transition $T$ in $x^{O,S}$ is $x_T^- = B^- u_T$. This part is removed through the firing of $T$ and results in a token $x_T^+ = B^+ u_T$. This token is moved through the independent part of the PN and results in a steady state vector $A^n B^+ u_T$ once the production process is completed. The vector $\bar{x}$ remains unchanged as long as no other transition is fired, which can be assumed without loss of generality as discussed in the poof of Theorem 1. Due to linearity of the cost function, the cost resulting from the unchanged part $\bar{x}$ coincides in the state $x^{O,S}$ and in $x^{O,C}$, which is why condition (4.18) reduces to

$$c_x^\top B^- u_T > c_x^\top A^n B^+ u_T. \tag{4.19}$$

If this condition holds for at least one input $u_T \in \mathbb{U}_{O,S}$, Theorem 1 and Corollary 1 can be used to conclude the existence of the sufficiently long prediction horizons $N_O$ and $N_O^+$ for operation $O$, respectively.

For a linear cost function, also the production cost $c_P(M, O)$ can be explicitly considered in the computation of a sufficiently long prediction horizon. This is possible since the cost of the common part $\bar{x}$ of the vectors $x^{O,S}$ and $x^{O,C}$ is independent of the cost for executing the operation $O$. If condition (4.19) is fulfilled for at least one $u_T \in \mathbb{U}_{O,S}$, a conservative upper bound on the sufficiently long prediction horizon $N_O$ can be computed with an algorithm described in [86, Algorithm 1]. This algorithm simply computes for all input vectors $u_T$ satisfying condition (4.19) the smallest value for the prediction horizon $N$, for which applying the input $u_T$ and then $u = 0$ for $N$ time steps yields any benefit.

It can be concluded that determining the existence of a sufficiently long prediction horizons $N_O$ and $N_O^+$ is simplified if a linear cost function is used. It can be computed by an algorithm described in [86, Algorithm 1], in which the production cost of executing the operation $O$ is explicitly considered. As before, with the values for $N_O$ and $N_O^+$ for all operations $O \in \mathcal{O}$, sufficiently long prediction horizons $N_J$ and $N_J^+$ for all jobs $J \in \mathcal{J}$ and ultimately sufficiently long prediction horizon $N$ and $N^+$ for the entire production problem can be computed.

## 4.4.4 Alternative Techniques for Guaranteed Convergence

The conditions in Theorem 1 and Corollaries 1 - 3 are only sufficient conditions for the completion of the operations, jobs, and the entire production problem, respectively. This means that there might also be the possibility that the production problem will be completed by the MPC schemes (4.2) and (4.12) in case condition (4.4) or (4.14), respectively, is not satisfied for all operations. This would, for example, be the case if the execution of a single operation $O$ cannot directly yield profit, and therefore does not satisfy condition (4.4), but it enables another operation $O'$ and the immediate execution of $O'$ after $O$ satisfies a condition similar to (4.4), namely $c(x^{O,S}, 0) > c(x^{O',C}, 0)$. Such cases could be captured by also investigating sequences of operations with respect to condition (4.4) in Theorem 1 instead of only single operations. This, however, will further increase the computational load to determine whether the completion of the production can be guaranteed for a given problem. It would potentially increase the length of the sufficiently long prediction horizon $N_O$ due to the larger production cost of two operations with respect to the production cost of a single operation. In the MPC scheme with terminal cost (4.12), it would require the prediction horizon $N$ to be at least as long as the production time of the first operation $O$, such that the second operation $O'$ can be actively started. Therefore, Corollary 1 does not hold any more for all $N \in \mathbb{N}_{>0}$ and the minimum computational load of the MPC with terminal cost increases.

In Theorem 1 and Corollaries 1 - 3, sufficiently long prediction horizons $N_O$, $N_O^+$, $N_J$, $N_J^+$, $N_{\min}$ and $N_{\min}^+$ are computed in order to guarantee completion of the respective elements of the production problem. An alternative approach, that renders the search for sufficiently long prediction horizons unnecessary, is applying an algorithm similar to the one presented by Scokaert and Rawlings [72, Algorithm 1] to determine a sufficiently long prediction horizon during runtime. It achieves closed loop stability and constraint satisfaction by first formulating a standard MPC problem similar to (4.2). At every time instant, starting from a given minimum value for the prediction horizon $N = N_0$, the MPC Problem (4.2) is repeatedly solved until its optimum is reached and $N$ is increased in every repetition. This procedure is stopped once the final state $x(N|k)$ of the optimal solution to Problem (4.2) reaches a terminal set $\mathbb{X}_f$. The stability guarantee is achieved by exploiting that the optimal solution of the MPC problem for all states inside the terminal set $\mathbb{X}_f$ is known to be stabilizing. The main idea that we can adopt from this approach is to increase the prediction horizon $N$ until the optimal solution of the MPC problem fulfills some specific properties, for which we know that they lead to the desired convergence. For the JMPMSF modeled as PN, those properties can be deduced from Lemma 1 based on Assumptions 1 and 2. In particular, such a condition can be based on the same arguments as the proof of Theorem 1 requiring that the optimal first input initiates a new operation, i.e., $u^*(0|k) \neq 0$. If a process is running, the prediction horizon can be arbitrary, but once a steady state is reached, the optimal first input $u^*(0|k)$ needs to be nonzero. Then, due to Properties (P2), (P6) and (P8) and Assumption 1, it follows that all operations will be executed. The condition that the optimal first input $u^*(0|k)$ is nonzero can be used to terminate the increase of the prediction horizon $N$ in the search for a solution with guaranteed convergence in closed loop. Due to the computational efficiency of the MPC formulation (4.12) even for arbitrarily long extended prediction horizons $N^+$, this approach is not expected to have much practical relevance.

It is known that increasing the prediction horizon $N$ generally leads to an increased region of attraction of standard MPC formulations, but that this has the drawback of an increased computational burden [46]. In the same way, the increase of the prediction horizon $N$ in the MPC scheme for production scheduling (4.2) leads to an increased region of attraction, meaning that more operations can be proven to be completed by the MPC scheme, at the cost of an increased computational burden. For the MPC scheme with an extended horizon (4.12) through the terminal cost $V_f$, an enlarged region of attraction can be achieved by increasing the extended horizon $N^+$, which does not affect the computational effort to solve the MPC problem. This is the case since increasing the extended horizon $N^+$ only changes the terminal cost $V_f$ in (4.13), but leaves the prediction horizon $N$ in the optimization problem (4.12) unchanged. In [46], the weight of the terminal cost is adjusted to increase the region of attraction without changing the computational load. Limon et al. [46] justify removing a stabilizing terminal set by increasing the weight of a terminal cost $V_f(x(N|k))$ until the terminal constraint is not active any more in the desired region of attraction. This is similar to the approach exploited in the MPC scheme with terminal cost (4.12), where the weight of the terminal cost is increased by increasing the length of the extended prediction horizon $N^+$. In [46], a larger weight on the terminal cost increases the region of attraction. This, however, makes the MPC scheme without terminal constraint perform worse in the nominal closed loop with respect to the provided stage cost function $c(x, u)$. Extending the prediction horizons $N$ and $N^+$ too much in the presented MPC schemes might also deteriorate the closed loop performance, since the effect of the terminal state dominates the production cost, as mentioned in the proof of Theorem 1.

In contrast to many approaches with terminal conditions from literature [67], enforcing the completion of the production problem cannot simply be achieved through a constraint on the terminal state $x(N|k) \in \mathbb{X}_f$ in the MPC problems considered in this thesis. In the approaches with terminal set constraints, the desired guarantees are achieved through hard constraints while the cost function mainly serves to pursue the economic objective. For the PN model of the JMPMSF with the goal to complete all operations, this approach does not lead to the desired convergence guarantees. Despite the terminal constraint, the MPC scheme could still favor a "lazy" solution, which plans to start the next operation at some time in the future and not to do anything in the current time step. If the terminal set is kept constant all the times the MPC problem is evaluated and the system is at a steady state, the same "lazy" solution will be the optimal one at every time step. The terminal set is repeatedly shifted further into the future and no operation is started. Only if the terminal constraint is fixed with respect to the real time and not with respect to the predicted time, the "lazy" solution needs to become active at some time and an operation is started. In this case, however, the robustness with respect to time delays in the system is lost, since the start of the next operation is artificially delayed in order to strive for the optimal solution. If some delay happens due to an unexpected disturbance, the system might become infeasible since it is not possible any more to reach the terminal set in the remaining time window.

Guaranteeing constraint satisfaction and closed loop convergence to a desires steady state through an appropriate cost term, as it is done in Theorem 1 and Corollary 1, is known in MPC literature. For example Feller and Ebenbauer [20] introduce a penalty term in the form of a relaxed recentered barrier function that penalizes being close to the boundary of a terminal set. They are able to prove strict constraint satisfaction and asymptotic stability of the origin by appropriately parametrizing the penalty on the terminal state $x(N|k)$. In this approach, the main advantages of moving the constraints to the cost function are improved feasibility conditions and computational efficiency since the resulting MPC scheme only requires to solve an unconstrained optimization problem. This method, however, requires a convex cost function, which is not required in the setup presented in this thesis.

## 4.5 Summary

On the basis of the well structured models resulting from the algorithms developed in Chapter 3, we formulated the scheduling of FMSs as an MPC problem in Section 4.1. Through the systematic design of the problem description, the mathematical model in the form of a PN and the resulting MPC problem, we were able to prove the completion of the production problem and its recursive feasibility. Based on the insights form the convergence proof of the MPC scheme (4.2), we derived a computationally more efficient MPC scheme with terminal cost (4.12) in Section 4.2. The optimization problems (4.2) and (4.12) are MPC formulations for the scheduling of a production problem given as a JMPMSF provided in the form described in Section 3.2 and modeled as PN through the algorithms introduced in Section 3.3. In Section 4.4, we discussed how the MPC schemes react to disturbances in the system and how the economic cost terms can be included in the cost function of the underlying optimization problem through an appropriate assignment to the PN elements. We discussed how the computation of a sufficiently long prediction horizons $N$ and $N^+$ becomes easier in the case of a linear cost function. Finally, in Section 4.4.4, we related the provided convergence results to results from MPC literature.

# Chapter 5

# Numerical Examples

In order to illustrate some features and capabilities of the model predictive control (MPC) scheme with terminal cost presented in Section 4, we simulate two scheduling problems from the literature. From the set of problems used by Lunardi et al. [51] for evaluating their models of an online printing shop, we select the small problem "sops1.json" and the medium-sized problem "mops1.json", which can be found online at `https://willtl.github.io/ops`[1]. We do not consider all effects of the online printing shop investigated in [51], and only simulate the scheduling problems in the form of a job shop with multi-purpose machines and sequence flexibility (JMPMSF) as described in Section 3.2. The most significant differences are that we do not allow the simultaneous execution of multiple tasks of the same job by different machines and the overlapping of consecutive tasks, and that we do not consider planned times of unavailability of the machines. Nevertheless, those problems are well suited to deduce JMPMSF descriptions and apply the MPC based scheduling scheme, since they represent production scenarios that are closely related to the considered flexible manufacturing systems. The scheduling problems describe scenarios of a complex online printing shop, in which mass customization is used to attract customers and reduce production cost. In an online system, customer specific products of different kinds are ordered and result in jobs with different sets of operations and arbitrary precedence constraints. The jobs have to be scheduled in a manufacturing system with machines that can execute various tasks. The resulting route through the system is not predefined but determined according to the requirements of the jobs and the capabilities of the machines.

To arrive at the JMPMSF description, the JSON files are parsed with a Matlab script and the relevant information is extracted. This parser is slightly different than the one used in [88], which is why the resulting JMPMSF descriptions slightly differ and the application of the MPC based scheduling scheme results in different closed loop schedules compared to [88]. The conclusions drawn from those examples, however, are valid regardless of the parser, since only the problem instances are different while their main characteristics remain unchanged and most significantly the application of the MPC based scheduling scheme reveals the same insights. Therefore, this section is based on and taken in parts literally from [88]. We first investigate the nominal

---

[1]accessed on November 06, 2021

case, in which the machines are always available, in Section 5.1. Then, we show the ability of the proposed scheduling scheme to handle unplanned machine breakdowns in Section 5.2.

# 5.1 Nominal Case

From the original problem description provided by Lunardi et al. [51], we consider the resource constraints (R5) and (R6), meaning that not all machines can execute every task and that every machine can only execute one task at a time. The precedence constraints among the tasks of the same job (R1) and the production times $t_P(M, \tau)$, which are already provided as a number of production steps $k_P(M, \tau)$, are specified in the JSON files. The Restrictions (R2) and (R3) requiring that tasks can only depend on other tasks in the same job and two tasks must not be mutually dependent, are already respected in the problem formulation. In the setup described in [51], preemption is not allowed and therefore Restriction (R4) is fulfilled. The single exception from this rule is the unavailability of machines, which is only considered in Section 5.2. The main difference to the JMPMSF proposed in this thesis is that in the original problem description, it is possible that multiple tasks of one job can be executed at a time, which violates Restriction (R7).

From the problem descriptions deduced from the JSON files, the Petri net descriptions are generated with Algorithms 3 and 4 implemented in Matlab, and their state space descriptions are used to formulate the MPC problems in the form described in (4.12). For the simulation of the manufacturing systems with MPC in closed loop, we use Matlab R2019b on a ThinkPad L390 Yoga with an Intel® Core™i5-8265U CPU. We do not take any measures to reduce the computation time, such as parallelization of independent computations. As a solver for the mixed integer linear program (MILP) (4.12) we use `intlinprog`, which is a standard solver for MILPs in Matlab.

The problems are initialized with all tasks of all jobs being necessary and none of them being completed. As no cost function is provided in the JSON file, we choose a linear cost function $c(x, u) = c_x^\top x + c_u^\top u$ that weighs every token in a starting place with 2, every token in a production place with 5, every token in a buffer place with 1, every token in a necessity place with 1, and every token in a completion place with 0. The cost of firing a transition is 1 for all transitions. This cost function fulfills Condition (4.14) for every operation $O$ due to the cost difference between necessity places and completion places, as described in Section 4.4.2. In every starting state $x^{O,S}$, a necessity place is marked with one token, and in every resulting state $x^{O,C}$ that can be reached by executing the operation $O$ through firing a transition $u_T \in \mathbb{U}_{O,S}$, this token is moved to a completion place. This results in a cost decrease of $c(x^{O,S}, 0) - c(x^{O,C}, 0) = 1$ between the states $x^{O,S}$ and $x^{O,C}$. With Corollary 1, for every operation $O$ there exists an extended prediction horizon $N_O^+$ for which it is guaranteed to be completed. Therefore, if a sufficiently large extended prediction horizon $N^+$ is used, the MPC scheme (4.12) is guaranteed to complete the entire production problem

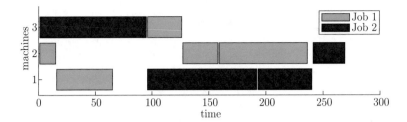

Figure 5.1: Gantt chart of the closed loop schedule for the small sized-scheduling problem "sops1.json" [51], resulting from an application of the MPC (4.12).

in closed loop according to Corollaries 2 and 3. To this end, the MPC scheme (4.12) is parametrized with the prediction horizon $N = 1$ and the rather long extended prediction horizon $N^+ = 500$.

The small-sized scheduling problem "sops1.json" has $n_M = 3$ machines, $n_\tau = 9$ tasks and $n_J = 2$ jobs. The MPC needs approximately 15 s to complete the scheduling problem "sops1.json" in closed loop and the resulting makespan of the entire production problem amounts to 269 time steps. During simulation, according to the standard MPC Algorithm 1, at every time step $k$ the MPC problem (4.12) is solved, the first part of the optimal input $u^*(\cdot|k)$, i.e., $u^*(0|k)$, is applied to the system. Then, the model of the production system is simulated one step further as the production would be continued in a real-world scenario. The resulting average computation time to solve a single MPC optimization problem is slightly less than 60 ms for the "sops1.json" with a maximum computation time of approximately 0.4 s. The Gantt chart of the resulting closed loop schedule is given in Figure 5.1.

The medium-sized scheduling problem "mops1.json" comprises $n_M = 8$ machines, $n_\tau = 39$ tasks and $n_J = 5$ jobs. It takes approximately 29 min to simulate the MPC in closed loop for the 592 time steps that it needs to complete the production problem. Solving a single MPC problem takes roughly 2.9 s with a maximum computation time of 3.4 s. The Gantt chart of the closed loop schedule determined by the MPC is shown in Figure 5.2.

The simulation results are not directly comparable to the ones presented in [51], since we do not consider some characteristics of their setup, in particular the parallel execution of multiple tasks of the same job and the overlapping of subsequent tasks. Additionally, we consider a different optimization criterion and simulate the production system in closed loop, meaning that we solve many different optimization problems, whereas Lunardi et al. [51] determine one optimal schedule for the initial problem. This fundamentally different approach leads to a reduction of the computation time for the MPC solution when a short prediction horizon is used. In the simulated extreme case of a prediction horizon of $N = 1$, the optimization in each time step

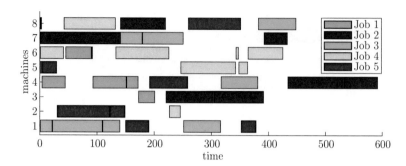

Figure 5.2: Gantt chart of the resulting closed loop schedule for the medium-sized scheduling problem "`mops1.json`" [51] when applying the MPC scheme (4.12).

of the "`mops1.json`" on a consumer laptop takes 2.9 s, whereas the solution of the entire scheduling problem on the "HPC facilities of the University of Luxembourg" takes 16.1 s [51]. This demonstrates the advantage that a computationally intensive problem, previously solved on a high performance computer, can be efficiently solved on a consumer laptop by applying the MPC based scheduling scheme.

Despite the fact that the cost function does not directly consider the makespan, the resulting closed loop schedule for the "`sops1.json`" is the shortest possible one for the considered problem. It is even shorter than the makespan determined by Lunardi et al. [51], which is only possible due to slightly different properties of the problem formulation. The shortness of the makespan illustrates how the production time is implicitly minimized under consideration of the given cost function as discussed at the end of Section 4.1.2. For the "`mops1.json`", the resulting makespan of the MPC solution is larger than the shortest one determined by Lunardi et al. [51], which completes the production problems in 344 time steps. However, their resulting schedule for the "`mops1.json`" is only possible due to parallel execution of multiple tasks and by partial overlapping among tasks of the same job, which we did not consider in our simulation. The resulting makespan of 592 time steps achieved by the MPC based scheduling scheme is only 14% longer than the shortest one possible for the "`mops1.json`" without overlapping and parallel execution of tasks. This small difference to the optimal makespan shows that although it is not directly minimized, a short makespan is still achieved.

The presented examples are not specifically designed to highlight the advantages of our approach to directly incorporate the economic objective of the scheduling problem, which is not given in the problem instances, and exploit the feedback mechanism to flexibly adapt to changes in the system. Nevertheless, they illustrate that the MPC is applicable to common scheduling problems and that it neither has a particularly long computation time nor leads to a bad solution in the closed loop. The implementation of the example shows another advantage of the presented framework. It allows to apply the MPC based scheduling scheme to solve scheduling problems with little effort. Only a simple parser was necessary to import the problem data and from then on the automatic model generation and control framework led to a solution of the problem. The capability of the presented scheduling scheme to flexibly adjust as a response to unplanned changes in the system is shown in the next section.

## 5.2 Machine Breakdowns

To illustrate the reactiveness of the proposed scheduling scheme, we repeat the simulation of the "mops1.json" scheduling problem, but with unplanned machine breakdowns. This is in contrast to Lunardi et al. [51], which can only cope with the unavailability of machines if it is known in advance. We simulate the problem with the same periods of unavailability as provided in the JSON file, but consider them to be breakdowns of the machines, which are not known in advance but detectable during runtime. For the simulation, we use the identical parameters as in Section 5.1, but at the times when the machines are unavailable, according to the problem description in the JSON file, the production processes executed on the unavailable machines are interrupted. If they do not execute any task at the time when a breakdown occurs, the machines become unavailable and no new task can be started on an unavailable machine. As described in Section 4.4.1, the interruption of a running process is realized by removing the token from the marked production place corresponding to the broken machine, which leaves the MPC problem feasible according to Lemma 2. For the restart of the production process once the machine is available again, the status of the interrupted process is restored. To this end, the token that is removed at the time when the machine breaks down is saved and it is restored once the machine becomes available again. Setting a machine which is currently not executing any task to unavailable is implemented similarly by removing the token from the idle place and restoring it once the machine is available again.

The results of the simulation with machine breakdowns are illustrated in Figure 5.3, where the periods in which the machines are unavailable are marked with grey bars. In comparison with the ideal simulation depicted in Figure 5.2, it can be seen that the MPC based scheduling scheme reacts to unavailable machines by assigning tasks to alternative machines. In the beginning, the results for the nominal case in Figure 5.2 and the scenario with breakdowns in Figure 5.3 are identical. The breakdowns of Machines 2, 4 and 7 between time steps 160 and 170 lead to delays in Jobs 1 and 3 and ultimately render the rescheduling of multiple jobs necessary. This also leads to rescheduling of Job 2, which is the critical job that is produced without any interruptions during the entire makespan. The machine breakdowns lead to a makespan increase of 83 time steps compared to the nominal case. The total delay on the critical

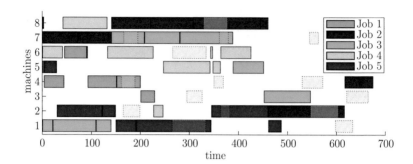

Figure 5.3: Gantt chart of the closed loop schedule for the medium sized scheduling problem "mops1.json" [51], resulting from the application of the MPC scheme (4.12) when unplanned machine breakdowns occur. The breakdowns are sown as gray boxes.

path is 104 time steps due to the breakdowns of the Machines 2 and 8. This example illustrates that feedback from the process can be integrated during production and that the MPC reacts to machine breakdowns with flexibly adjusted schedules. The instantaneous reaction mitigates the resulting delay in the completion of the production problem.

## 5.3 Summary

In this chapter, we performed three simulation examples to show how the MPC scheme with terminal cost introduced in Section 4.2 can be applied for the scheduling of flexible manufacturing systems. We considered exemplary scheduling problems from the literature to illustrate that the MPC scheme is applicable to typical problems in this field. The problems were simulated on an ordinary laptop, which did not need much time to solve the underlying optimization problems, which indicates that small- and medium-sized problems are computationally tractable. In the last example, machine breakdowns were considered and the MPC scheme reacted by rescheduling the tasks accordingly, which shows the flexibility of the presented approach.

# Chapter 6

# Conclusion

Recent developments in industrial production driven by the advent of the fourth industrial revolution require an increase of the flexibility of production plants. In the vision of Industry 4.0, the goal of enabling the production of customer specific products at the same cost as in mass production demands for structures and concepts that are able to react quickly to changes in the production and in the market. This not only affects the hardware setup of the plants, which become flexible manufacturing systems (FMSs), but also gives rise to new challenges for their coordination and control, which motivates the study of a novel flexible production scheduling framework in this thesis. More precisely, the developed framework covers the whole range from the problem formulation over the deduction of a mathematical model until the solution of the scheduling problem for the FMS. In the context of Industry 4.0, the concept of skill-based engineering proposes a modular factory design for which already approaches for the design of its digital structure exist. In this digital environment, the resulting digital twins of the components hold the relevant information on the real production system. To draw benefit from those structures, the provided flexibility needs to be exploited, as is the case in the proposed framework. The research question considered in this work was how to make use of the flexibility of modern manufacturing system with an automated modeling and control framework.

The first part of this thesis addressed the problem formulation and the automatic generation of a mathematical model. In particular, the scheduling for the production of customer specific goods in an FMS is considered as a class of scheduling problems, which provides many relevant types of flexibility. From this problem formulation, a Petri net (PN) model is automatically generated which is particularly suited for the application of model predictive control (MPC) for the scheduling of the production, due to its algebraic state space representation.

In the second part of this thesis, based on this particular type of models, two MPC schemes are proposed for production scheduling. They exploit the properties of the model in order to provide theoretic guarantees on the closed loop resulting from applying the MPC to the manufacturing system. In particular, we prove that the MPC optimization problem is recursively feasible and that the application of the resulting optimal inputs complete the production problem in closed loop while taking an economic objective into account.

To summarize, by explicitly taking into account the characteristics of a skill-based engineering approach, we contributed to the production scheduling for FMS. Introducing an integrated framework that considers modeling as well as scheduling of an FMS, is considered an important step for realizing mass production at lot size one.

In the following sections, we summarize our results in more detail, discuss their advantages and limitations, and give an outlook on possible future research directions based on the results achieved in this thesis. Parts of the following discussion and the outlook on future work is based on the discussion in [88] and taken in parts literally from it.

# 6.1 Summary

In Chapter 2, we provided the required background for the development of the reactive scheduling scheme, which covers the four main fields related to the developed framework. At first, we explained the theoretical concepts envisioned in Industry 4.0. In this context, we mainly focused on the concepts of the digital twin and of skill-based programming, which point out the need for a modular and flexible scheduling scheme. As the second field covered in this thesis, we gave a brief introduction in production scheduling and explained the most relevant terms. For being able to classify the problem class considered in the remainder of the thesis, we introduced different types of flexibilities, which can be available in scheduling problems, and a common classification scheme for production scheduling problems. After a brief introduction to the theory of Petri nets, we explained the fundamentals of MPC on which the proposed scheduling scheme is based.

In Chapter 3, we started by describing the overall framework for the reactive scheduling of FMSs and explaining the relation between its different parts. On the one hand, this framework includes the automatic generation of a mathematical model for the manufacturing system. On the other hand it considers the application of the generated model for the online scheduling of the manufacturing system in a feedback control loop. An advantage of the proposed framework is to provide the automated modeling as well as the automatic control of the FMS based on a simple problem formulation. In this problem formulation, we described the FMSs as a particular class of scheduling problems, which, in accordance with a general classification scheme from literature, we classified as job shop with multi-purpose machines and sequence flexibility (JMPMSF). This type of scheduling problems was introduced as the basis for the reactive scheduling of FMSs, since it offers the required flexibilities to describe a modular FMS based on skill-based engineering in an Industry 4.0 context and allows to deduce various more specific scheduling problems as special cases. To this end, the results provided for the JMPMSF can be directly applied to special cases that are relevant in many applications, as we laid out in Section 3.2.2. We showed how the machines and products being produced in the FMS are assigned to one another in a scheduling problem described in the modular formulation of the JMPMSF.

Enabled by the modularity of the problem description, we introduced automatic model generation algorithms in Section 3.3, which result in a Petri net model as a discrete time representation of the FMS. The goal of the model generation was to create a mathematical representation of the JMPMSF that is suited to be used for MPC and allows to provide rigorous guarantees by system theoretic analysis. To this end, we slightly enhanced the classical notion of PNs and separated an automatically executed part from an actively controlled one. Through this distinction, we were able to formulate two algorithms, which together lead to a PN model that describes the JMPMSF in a discrete time representation. In the resulting model, the dependencies and restrictions between the elements in the JMPMSF are captured by the graph structure of the PN, while the decisions in the scheduling problem are realized by means of the actively controlled transitions. The algebraic description of the PN is a linear discrete time system, in which the inputs represent the firing of the transitions. The graph structure represents the dependencies and restrictions and results in the system matrices and a liner inequality constraint, which needs to be met such that no negative tokens are generated. The linear system together with the inequality constraint is called state space description of the PN. We analyzed the mathematical model and showed that the characteristics of the JMPMSF, in particular with respect to its flexibilities and restrictions, are preserved through the transformation process. As preparation for the MPC schemes, we described the properties of the PN model that are important to formulate the control schemes for the manufacturing system and to guarantee their characteristics.

In Chapter 4 we exploit the simplicity of the algebraic description of the PN to derive a feedback control law in terms of MPC. The optimization problem to solve in each time step is a discrete time MPC formulation. It considers the linear system dynamics, the linear constraint to ensures that no negative tokens are created in the PN, and an integer constraint on the input which ensures integer-valued tokens. Based on the specific properties of the scheduling problem captured in the PN model, we were able to prove recursive feasibility of the MPC control law and to formulate a sufficient condition to guarantee the completion of the scheduling problem. The guarantees rely on mild assumptions on the problem formulation and on a criterion for the cost function employed in the MPC optimization problem. To reduce the computational effort that is required to solve the initial MPC problem, we propose a more efficient formulation based on a terminal cost. Also for this formulation, the completion of the production problem is guaranteed under the same prerequisites. Since the decrease condition required to ensure the completion of the scheduling problem is hard to verify in general, we elaborated the simplifications that are possible in the case of a linear cost function. In an analysis of the cost function with respect to the original problem formulation, we related the MPC formulation to the manufacturing problem. In this regard, we discussed how the manufacturing cost can be transferred from the original problem formulation into the cost function in the MPC problem. By that, the economic objective for the scheduling problem of the FMS is explicitly considered in the reactive scheduling scheme.

Finally, in Chapter 5, we showed the applicability of the MPC based scheduling scheme at a small and a medium sized example taken from literature. The simulations revealed that the MPC formulation with terminal cost is computationally tractable and leads to satisfactory results. When machine breakdowns occur, the scheduling scheme exploits the flexibility of the JMPMSF and leads to an adjusted schedule in the closed loop.

# 6.2 Discussion and Outlook

We introduced a novel reactive scheduling framework for FMSs in this thesis. It builds upon a modular problem formulation in the form of a JMPMSF. For its formulation we assumed the implementation of concepts discussed in Industry 4.0, particularly of the digital twin and skill-based engineering concepts. This assumption is not yet completely fulfilled. A required preliminary for the application of the proposed reactive scheduling scheme in practice is a standardized skill taxonomy and a uniform level of abstraction to unify how skills and their resulting tasks in the JMPMSF are defined such that they can consistently be matched [54]. This needs to be considered on both sides, i.e., on the product side and on the side of the machines. The need for defining standards even goes one step further to the hardware level, where a physical realization of the skills is required [62]. The production plans of the products as well as the composition of the manufacturing system need to be compatible with the definition of the sills. To determine the required skills for the manufacturing of a product and the assured skills of the different machines, expert knowledge is required [69]. This limitation holds for all automated scheduling schemes that rely on the abstraction of physical properties and capabilities. Only an expert in the field of application can define reasonable skills and determine reasonable cost functions to be minimized in the MPC based scheduling scheme.

**Extensions in the Problem Formulation**

Compared with other scheduling schemes, we only consider a few of the possible restrictions in the scheduling problem. The most common more restrictive scheduling problems, as for example the job shop, the flow shop or the open shop, can be directly considered in the presented setup as we explained in Sections 3.2.2 and 3.3.3. However, more specific restrictions and characteristics of particular scheduling problems require further considerations. For example the online printing shop considered by Lunardi et al. [51] is subject to specific challenges, which we did not address and therefore also did not consider in the examples in Section 5. Also Brucker [10] describes different job characteristics, which we have not considered explicitly. As stated in Section 3.3.3, however, the consideration of further properties and constraints of a manufacturing system is possible in the presented framework, but requires adaptations of the employed algorithms. As an example, the explicit handling of waste can be considered through dedicated PN structures in which waste products are generated during production and stored in specific buffers that need to be emptied from time to time. Also, assembly tasks connecting multiple parts that can be handled separately before an assembly step, but constitute one combined intermediate product afterwards, can be introduced. The structure in which the intermediate parts have to be combined can be represented in a dependency graph between separate workpieces [5]. The assembly of two workpieces can be implemented as the combination of the results of two jobs. In the PN model, this can be realized as combining the parts which are stored in two buffer places that are marked after the last task in which the intermediate products are completed by means of a synchronization transition.

Transportation times can be considered by extending the chain of production places by the length that the transportation towards the next machine needs in the worst case. This leads to a chain of production places that represent the combined transportation and production process. To implement different transportation times in different cases, not all starting transitions move the token to the first place of this chain. Depending on the actual transportation time required to move the product to the respective production place, the starting transitions move it to a place that is already later in the combined chain of the transportation and production process. By this mechanism sequence dependent transportation times can be realized. The same mechanism can also be used to implement sequence dependent setup times, which are a relevant property of manufacturing scenarios described in literature [13, 15, 51].

The minimization of the makespan, which is the most common scheduling objective in literature [13], was not explicitly discussed in this thesis. Also hard deadlines for the completion of the jobs were not considered. Since in practice it is important to adhere to deadlines and it might be beneficial to produce as fast as possible, those are relevant future research topics to be considered in the presented framework. Minimizing the makespan can be achieved by appropriate cost functions that assign an additional weight to uncompleted tasks and jobs at time instants further in the future. As discussed in Section 4.4.2, this weighting is even beneficial with respect to the condition (4.4) used for guaranteeing the completion of the operations. Under which conditions such a weight will actually lead to the minimization of the makespan in closed loop, however, is not obvious and deserves further investigation. Including hard deadlines as time dependent state constraints in the MPC problem in a straight forward manner is possible, but it threatens the recursive feasibility of the MPC problem. Therefore, further research is required in this case and tools from MPC theory, for example in combinations with a terminal set constraint, are promising in this regard.

Due to the transformation of the JMPMSF into a linear discrete time system with a linear inequality constraint, the resulting state space description has a common form for MPC formulations. Therefore, it offers the possibility to easily include the dynamics of other systems that can be expressed or approximated by linear discrete time systems. As an example, the interface between a continuous process and the discrete valued PN model can be realized by a controlled transition. The continuous process evolves over time according to its discrete time system description without any underlying PN structure. Therefore, it can attain arbitrary non-integer values. The interfacing controlled transition can withdraw a certain amount from the continuous process and generates a token in its output place. Whether it is allowed to fire might depend on the continuous valued process and can be formulated by an inequality similar to the non-negativity constraint (2.3) without requiring changes of the adjacent PN model or the MPC formulation. Such structures allow a combination of different types of systems, for example at the interface between a continuous chemical production and a packaging and transportation system that withdraws discrete value batches.

To summarize the previous paragraphs, discussing further possible extensions of the problem formulation in more detail, enhancing the presented algorithms and proving their resulting properties offers plenty of opportunities for future research. Finally, only applying the presented framework to real-world problems with relevant scheduling objectives will show its usefulness in practice.

**Further Analysis Possibilities**

On the basis of the Petri net description of the JMPMSF, we focused on analyzing its state space description and provided important properties in Lemma 1. This does not directly exploit results and analysis techniques for PNs, despite this field offering a wide range of methods that fill multiple books [16, 73]. Exploiting this branch of research to gain further insights in the problem, to explore other PN structures to model particular characteristics of a production system and to deduce other means to control the system holds plenty of options for further research.

In the presented approach, we focused on the solution of the scheduling problem by means of MPC. The most important features are the guaranteed feasibility of the employed optimization problem and the guaranteed solution of the entire scheduling problem in closed loop in the sense of convergence to a state in which all tasks of all jobs are completed. The guarantee for recursive feasibility is directly given by the problem formulation. For the guarantee of the completion of operations in Theorem 1, it is assumed that the optimization performed during computation of the MPC control law is solved until the optimum.

The MPC optimization problems (4.2) and (4.12) formulated in Chapter 4 are designed for the solution of the scheduling problem presented in Section 3.2, which is known to be NP-hard [15]. They are finite horizon approximations of the overall optimization problem that solves the whole scheduling problem until completion and therefore less computationally expensive. Nevertheless, their solution space still grows exponentially with the number of jobs and alternative machines. Thus, the computation of the optimal input trajectory $u^*(\cdot|k)$ is computationally demanding.

Although it is assumed that the optimization is solved until the optimum, the MPC optimization problem does not need to be solved completely in every time step. On the one hand, the last $N-1$ steps of the solution computed at time step $k$ can be used as the first $N-1$ steps of a warm start solution at time step $k+1$ as it is common in MPC [67]. As appending such a candidate solution with the input $u(N-1|k+1) = 0$

is always feasible as shown in the proof of Lemma 3, a feasible candidate solution of the optimization problem can be found instantaneously. On the other hand, also a suboptimal solution might be sufficient to complete the production problem. The optimal solution to the MPC problem is not needed for the guarantee of the completion of the production problem, which is already provided in Lemma 5 and based on a convergence property provided in Lemma 4. For the proofs of these Lemmas, the key criterion is that a non-zero input is applied to the system once it entered a steady state. If this property also holds for suboptimal solutions, the closed loop system will still converge to a state in which the production problem is completed. If we take this as an additional assumption or as an additional constraint in the optimization problem, the convergence guarantee provided for the optimal solution of the MPC optimization problem is retained also for suboptimal solutions. Therefore, if the optimal solution of the admittedly complex MPC optimization problems (4.2) and (4.12) cannot be found in the available amount of time, the MPC based scheduling approach is still a viable solution technique for the scheduling problem. If an optimization algorithm is used which is able to return a suboptimal but feasible solution after a desired amount of time, the MPC based scheduling approach provides the properties of a so called *anytime MPC* scheme. This means that it guarantees important properties of the closed loop system no matter how many iterations of the optimization algorithm are performed [21]. Additionally, the important arguments in the proofs of Theorem 1 and Corollaries 1 - 3 are based on steady states, which by definition persist over time. Therefore, it can also be argued that the optimizer has enough time in the important cases such that the optimal solution is attained and the properties can be guaranteed despite the high complexity of the optimization problem.

The high computational demand for the solution of the MPC problems with a long prediction horizon $N$ poses a limitation on the applicability of the proposed schemes. Motivated by the potential further increase in flexibility and the trend towards cyber-physical production systems, the modular structure of the problem setup and the deduced model can be further exploited. The model generated in Section 3.3 from the problem setup described in Section 3.2 consists of separate processes for each job that are linked via idle places of the machines. From this structure, an agent-based approach similar to the one presented by Xiang and Lee [89] can be developed, or the problem can be formulated in the form of a distributed MPC problem [52]. One possible agent-based approach would be to consider product agents for every job. The product agents then solve their local optimization problem and compete with each other for the machines. Also the converse setup is possible such that the machines compete for the jobs in an attempt to be productive and reduce the cost of open tasks that they would be able to execute. In a distributed MPC setup, the individual goals can be pursued by the individual agents while the resulting cooperative goal is negotiated online, as for example described in [42]. In our case, this requires coordination between the cyber-physical production systems in order to arrive at an optimal production schedule. Depending on the interaction mechanism, this leads to a distributed optimization problem that approximates the centralized optimization problems used in the MPC problems (4.2) and (4.12). Such solutions are favorable in Industry 4.0 scenarios due to their scalability [91]. The smaller individual optimization problems will be easier to solve. On the other hand, the global optimal solution might only be obtained in special cases and potentially only after multiple iterations between the competing agents [52]. At the end, there will be a compromise between the computational effort and the quality of a distributed solution. Especially if the cost function depends nonlinearly on cost terms related to different agents, as for example through a nonlinear dependency on the maximum power consumption of the plant, information on the whole system state would be required to find the optimal solution. Also constraints concerning multiple agents need to be carefully considered in the design of the communication and interaction network for the agents in order to guarantee an overall feasible solution [48].

With respect to computational load to solve the optimization problem in a setup that guarantees completion of the production problem, the formulation (4.12) with terminal cost is significantly more efficient than the formulation (4.2) without it. This is achieved by only requiring the optimization over a short horizon while still considering the effects over a long prediction horizon. In the initial formulation (4.2), the optimization window needs to be enlarged in order to guarantee convergence. In both presented MPC schemes, the closed loop performance may be improved by extending the prediction horizon $N$ under certain conditions. The introduction of the terminal cost and a longer extended prediction horizon $N^+$ in the MPC formulation (4.12) only changes the convergence guarantees, but not the expected performance of the MPC, since the optimization still only considers the prediction horizon $N$. Nevertheless, the MPC formulation with extended horizon (4.12) offers the possibility to increase the optimization horizon $N$ in an attempt to improve the closed loop performance without loosing any guarantees. The drawback of a large $N$ in both MPC schemes is that it leads to longer computation times due to increasing dimension of the optimization problems. Even though it is not required that the optimum is attained in the MPC optimization in order to guarantee convergence as discussed above, the optimal solution of the open loop optimization problem also promises better closed loop performance as a suboptimal solution. However, from MPC theory it is known that the repeated solution of a finite horizon optimal control problem does generally not lead to the infinite horizon optimal solution [28]. It is not clear whether the simple correlation between a long prediction horizon, which considers effects in the more distant future, and an increase in performance holds. It is not obvious which conditions on the cost function would be required for this simple correlation to hold. Due to the difference between open loop and closed loop performance, such conjectures are prone to be false [56]. Therefore, an analysis of the closed loop performance of the controlled system might yield interesting results. In this respect, it can be analyzed how the prediction horizon $N$ must be chosen to achieve some specified performance level with respect to the infinite horizon optimal solution, or which properties of the cost function lead to the optimal solution in closed loop when repeatedly performing a finite horizon optimization.

The special case of a linear cost function was already discussed in Section 4.4.3 and it was shown that exploiting linearity reduces the effort for providing the convergence guarantees with Theorem 1. In a similar way, further classes of cost functions, for example exhibiting certain separability conditions, can be analyzed. For the computation of a sufficiently long prediction horizon $N_{\min}$ that guarantees the completion of a given scheduling problem, specific algorithms could be formulated, which exploit the special characteristics of the cost function and thereby simplify the computation.

The feedback mechanism of the MPC schemes introduces a certain amount of robustness against disturbances and changes in the system, as we discussed in Section 4.4.1 and showed with an example in Section 5.2. However, as the growing demand for flexibility in Industry 4.0 might increases the uncertainty of the evolution of the manufacturing system, those effects might become even more relevant. Therefore, in future work the robustness of the closed loop system can be analyzed more rigorously, for example in terms of some assumed changes with respect to the initial manufacturing problem. Such changes might be machine dropouts or jobs that arrive during runtime and can be considered as disturbances. If the worst case disturbances are considered, methods from robust MPC can be applied, and if the disturbances are described probabilistically, the problem falls into the domain of stochastic MPC [67]. Especially with respect to machine dropouts, where the worst case scenarios are overly conservative, the consideration of a dropout probability is promising. In this case, stochastic MPC minimizes the expected cost in the presence of disturbances over the prediction horizon and thereby maximizes the expected profit in the manufacturing system.

With respect to the particular model of a flexible manufacturing system, more specific robustness metrics can be defined. In the problem description as a JMPMSF, the number of alternative machines that are capable of executing some task is a means to quantify the redundancy in the system with respect to this specific task. This constitutes a hard limitation on the number of machines that can concurrently break down before the problem becomes unsolvable. When analyzing the problem based on the PN model and with respect to the cost function, not all machines might be able to generate profit from executing all tasks that they are able to execute. On this level, a more specific redundancy metric can be formulated based on the completion states $x^{O,C}$ corresponding to each starting state $x^{O,S}$ of an operation $O$. Only pairs of starting and completion places which fulfill condition (4.4) can potentially be chosen by the MPC and they will only be chosen if the prediction horizon is sufficiently long. A robust MPC would then have to avoid starting states with a low number of corresponding completion states. Such a behavior allows more machine breakdowns before the problem becomes unfeasible and thereby is more robust with respect to machine breakdowns.

Another branch of future research can exploit methods from the field of MPC to tackle relevant problems in production scheduling. An example for this is the consideration of so called nervousness induced by reactive scheduling schemes, like the one presented in this thesis, which is considered to be one of the drawbacks of reactive scheduling [18, 43, 56]. Schedule nervousness of a scheduling scheme means that it tends to frequently change the input sequence due to external influences, which leads to discomfort of the operator. Oftentimes the cost of rescheduling is not appropriately considered and the resulting closed loop schedules are not optimal any more in practice. This problem can be tackled on the basis of the MPC formulation of the scheduling problem by using a penalty term on the change of the planned input sequence $u(\bar{k}|k) - u(\bar{k} + 1|k - 1)$. Preventing large or frequent changes in the input sequences by this means is well known in PN literature and has already been applied to production scheduling [56, 67]. For the case of the presented reactive scheduling scheme, the influence of such a penalty term needs to be rigorously analyzed, in particular with respect to its effect on the convergence guarantees of the MPC based scheduling scheme.

To summarize, further research related to the results in this thesis could consider

- the standardization in skill-based engineering as basis for the application of the presented framework,

- the integration of further characteristics of production systems in the model generation algorithms,

- developing a method to explicitly consider the minimization of the makespan and hard deadlines in the proposed scheduling scheme,

- the combination of the automatically generated model with other types of systems modeled as linear discrete time systems,

- analyzing the generated PN model with techniques from the PN literature,

- investigating computational aspects of the MPC schemes with respect to possible simplifications and the application as anytime MPC scheme,

- the formulation of the scheduling problem as an agent-based or distributed MPC scheme in the presented framework,

- determining closed loop performance guarantees for the proposed scheduling scheme,

- analyzing special classes of cost functions with respect to possible simplifications of the presented methods,

- applying robust and stochastic MPC techniques to the scheduling problem, and

- using analysis techniques from MPC theory to tackle related problems from the field of production scheduling in the presented framework.

In conclusion, the framework presented in this thesis can already be applied for the solution of a wide variety of relevant scheduling problems. It exploits the modularity of skill-based production systems in order to generate a simple model in the form of a linear discrete time system, which allows to directly apply MPC. The resulting reactive scheduling scheme is robust with respect to changes in the manufacturing system through the inherent feedback of MPC and adaptions based on the model generation algorithms. It is able to consider economic performance criteria while being proven to complete the scheduling problem in closed loop. Finally, the framework offers plenty of possibilities for further extensions and research opportunities.

# Appendix A

# Proof of the Properties of the Automatically Generated Petri Net

In Section 3.3, we stated some important properties of the Petri net and its state space description, which have been generated from the job shop with multi-purpose machines and sequence flexibility (JMPMSF) with the Algorithms 3 and 4. Those properties were formulated as Lemma 1. Since they are not directly apparent at first glance, we will now formally prove them based on the structure of the JMPMSF and the creation algorithms.

*Proof of Lemma 1 [88].*

(P1) We have to prove that there is no steady state $(x^s, u^s)$ with $u^s \neq 0, u^s \in \mathbb{N}^m$. As no entry in $u^s$ can be negative, the effect of a positive entry in $u^s$ through $B^+$ can only be compensated by the effect of another positive entry in $u^s$ through $B^-$, and vice versa. In the same sense the effect that an input arc $(P, T)$ has when the transition $T$ fires in the firing count vector $u^s$ can only be reversed by an output arc $(T', P)$ of a transition $T'$ in $u^s$. Thus, the claim is verified by observing that only starting transitions $T_{(...,S)}$ are handled in the controlled part of the PN $\mathbb{T}_C$. Every starting transition has an input arc $(P_{(...,N)}, T_{(...,S)})$ from a necessity place $P_{(0,\cdot,\cdot,N)}$ and its effect cannot be reverted, since there is no arc $(T', P_{(...,N)})$ back to any necessity places from any other transition.

(P2) For this property, the independent part $\mathbb{T}_I$ of the PN has to be considered, as it influences the matrix $A$ which is relevant to characterize steady states. A state $x^s$ is a steady state, if and only if no place $P$ is marked that is an input place of a transition $T \in \mathbb{T}_I$; otherwise, the automatic firing of the transition $T$ would remove the tokens from $P$ according to the arc $(P, T)$. (It is assumed that there is no transition which only has a trivial loop from a single place $P$ back to the same place as such a transition does not have any effect.) Since all production transitions $T_{(...,P)}$ and finishing transitions $T_{(...,F)}$ are handled in the independent part, none of their input places is allowed to be marked in a steady state $x^s$. As every production place $P_{(...,P)}$ is either input place of a production- or finishing transition, none of them is allowed be marked in a steady state. By the Property (I1) of independent transitions, the reverse direction is true as well.

(P3) The initial state $x^0$ is a vector of natural numbers, i.e., $x^0 \in \mathbb{N}^n$, since in Algorithm 3 the places are only marked with one token or they are not marked at all. That $x^0$ is a steady state follows directly from Property (P2) and by noticing that in Algorithm 3 no production place is marked.

(P4) Follows from the fact that $A \in \mathbb{N}^{n \times n}$ and simple algebraic arguments.

(P5) If no production place is marked in the precondition $x^s = B^- u$, it follows from Property (P2) that $x^s$ is a steady state. The preconditions of the controlled transitions $T \in \mathbb{T}_C$ constitute their input places. As only starting transitions $T_{(\ldots,S)}$ are in the set $\mathbb{T}_C$, only the input places of the starting transitions need to be considered. Due to Property (P2) and since no production place $P_{(\ldots,P)}$ is an input place to a starting transition created in Lines 15 and 16 of Algorithm 4, the Property (P5) is true.

(P6) An operation $O = (\tau, J)$ is completed if its completion place $P_{(0,\tau,J,C)}$ is marked. A completion place is only input place to starting transitions. If a completion place is input place to some starting transitions $T_{(\ldots,S)}$ through an arc created in Line 18 of Algorithm 4, it is also one of its output places through an arc created in the same line and the weight of both arcs is the same. Therefore, the removed tokens are immediately returned. In Algorithm 4, no input transition is created to the necessity places. Therefore, they remain unmarked once they had been unmarked by the firing of one of their output transitions.

(P7) This follows directly from $A$ being a matrix of natural numbers, i.e., $A \in \mathbb{N}^{n \times n}$.

(P8) First note that all finishing transitions $T_{(\ldots,F)}$ mark an idle place $P_{(\ldots,I)}$, a buffer place $P_{(\ldots,B)}$ and a completion place $P_{(\ldots,C)}$. All of those places only have starting transitions $T_{(\ldots,S)}$ as output transitions, which are all controlled, i.e., $T_{(\ldots,S)} \in \mathbb{T}_C$. Thus, every finishing transition ends an independent firing sequence leading to it. Every production sequence terminates in a steady state, as every production sequence generated in the Algorithms 3 and 4 ultimately ends with a finishing transition after $k_P$ steps. If no controlled transition $T \in \mathbb{T}_C$ is fired, no new production process is started and thus a steady state $x^s$ is reached.

In the matrix $A$, a production process is represented as a shifting sequence which ends in a state $\bar{x} = A^{\bar{k}} x$, in which and entry $\bar{x}_i$ with $x_i^{ID} = (M, \tau, J, P_{k_P(M,\tau)})$ is marked. Multiplying $\bar{x}$ with $A$, the entry $\bar{x}_i$ is multiplied with the i-th column of $A$, which corresponds to the finishing transition $T_{(\ldots,F)}$ created in Lines 10 and 11 of Algorithm 4 being the output transition of $P_{(M,\tau,J,P_{k_P(M,\tau)})}$, as described in Section 3.3.1. After this multiplications a state $x^s$ is reached in which only entries corresponding to places without independent output transitions are marked, i.e., idle places $P_{(\ldots,I)}$, buffer places $P_{(\ldots,B)}$ and completion places $P_{(\ldots,C)}$.

The diagonal entries corresponding to those places are left unchanged and no entries are added to the respective columns in the process of generating the matrix $A$ form the PN. Multiplying the resulting vector $x^{\mathrm{s}}$ with $A$ once again does not change those entries in $x^{\mathrm{s}}$ any more.

This holds for all production sequences as they are all generated with Algorithms 3 and 4 and thus the system ultimately enters a steady state with $x^{\mathrm{s}} = Ax^{\mathrm{s}}$. Since no production sequence can be larger than the dimension of the matrix $A$, i.e., $k_{\mathrm{P}}(M, \tau) \leq n$ for all $M \in \mathcal{M}$ and $\tau \in \mathcal{T}_M$, the steady state is reached in the worst case after $n$ time steps and the steady state can be computed with $x^{\mathrm{s}} = A^n x$.

(P9) To prove this property we have to analyze the input places that correspond to the starting transitions $T_{(\cdot, \tau', J, \mathrm{S})}$ of the investigated operation $O = (\tau', J)$. Those all have the necessity place $P_{(0, \tau', J, \mathrm{N})}$ and the completion places $P_{(0, \tau'', J, \mathrm{C})}$ of all $\tau'' \in \mathcal{T}_{\tau'}$ as common input places. $P_{(0, \tau', J, \mathrm{N})}$ will remain marked until one of the starting transition was fired and task $\tau'$ was executed and the places $P_{(0, \tau'', J, \mathrm{C})}$ will remain marked due to property (P6). For each of the investigated starting transitions, their further input places are one idle place $P_{(\cdot, 0, 0, \mathrm{I})}$ and one buffer place $P_{(\cdot, \tau, J, \mathrm{B})}$ of another task $\tau$ of the same job $J$. In every steady state, the idle places of all machines $P_{(\cdot, 0, 0, \mathrm{I})}$ are marked and thus this holds for the steady states $x^{\mathrm{s}, 1}$ and $x^{\mathrm{s}, 2}$. This is the case as only starting transitions remove tokens from the idle places and they inject them into a production sequence that ultimately leads back to the same idle place from which the token was removed. This happens through a completely independent sequence, i.e., a sequence which is completely represented in the matrix $A$ (cf. Figure 3.5).

Now it needs to be shown that, in $x^{\mathrm{s}, 2}$, at least one buffer place $P_{(\cdot, \tau, J, \mathrm{B})}$ is marked that has a starting transition $T_{(\cdot, \cdot, \tau, \tau', J, \mathrm{S})}$ of the investigated task $\tau'$ as output transition, except when the task $\tau'$ already had been executed. It is known that one such buffer place was marked in the steady state $x^{\mathrm{s}, 1}$, since otherwise $\tau'$ could not have been started. This means that for the pair of tasks $\tau, \tau'$ the condition in Line 13 of Algorithm 4 was true and for every combination of machines $M, M'$ the starting transition $T_{(M, M', \tau, \tau', J, \mathrm{S})}$ with the input buffer $P_{(M, \tau, J, \mathrm{B})}$ is created.

It remains to be shown that, if the task $\tau'$ was not already executed, starting from the buffer place $P_{(\cdot, \tau, J, \mathrm{B})}$ only other buffer place $P_{(\cdot, \bar{\tau}, J, \mathrm{B})}$ can be marked in any steady state $x^{\mathrm{s}, 2}$, for which the condition in Line 13 holds as well for the pair of tasks $\bar{\tau}, \tau'$ and thus the transition $T_{(\cdot, \cdot, \bar{\tau}, \tau', J, \mathrm{S})}$ is created to start $\tau'$. Such a buffer place $P_{(\cdot, \bar{\tau}, J, \mathrm{B})}$ can only be marked through a sequence that started with a starting transition $T_{(\cdot, \cdot, \tau, \bar{\tau}, J, \mathrm{S})}$, that was created in Line 15 of Algorithm 4. Thus, for the pair $\tau, \bar{\tau}$ the condition in Line 13 was true. If it is also true for the pair $\bar{\tau}, \tau'$, we found the buffer place $P_{(\cdot, \bar{\tau}, J, \mathrm{B})}$ as the required one to start the task $\tau'$ form $x^{\mathrm{s}, 2}$ through a transition $T_{(\cdot, \cdot, \bar{\tau}, \tau', J, \mathrm{S})}$.

Let us investigate the condition in Line 13 for this pair step by step. The first statement in this condition ($\bar{\tau} \neq \tau'$) is true since otherwise the task $\tau'$ was just finished. The second statement in this condition ($\tau' \notin \bar{\mathcal{T}}_{\bar{\tau}}$) is true since otherwise the previous task $\bar{\tau}$ could not have been started, since the task $\tau'$ was not yet completed. The last statement in this condition ($\{\tau^* \in \mathcal{T}_J : \bar{\tau} \in \bar{\mathcal{T}}_{\tau^*}, \tau^* \in \bar{\mathcal{T}}_{\tau'}\} = \emptyset$) is true as otherwise such a task $\tau^*$ would not have allowed task $\tau'$ to be started in $x^{s,1}$, which we know was the case as stated in the precondition of Property (P9). This concludes the proof.

$\square$

# Bibliography

[1] M. Alamir and G. Bornard. "Stability of a truncated infinite constrained receding horizon scheme: the general discrete nonlinear case." In: *Automatica* 31.9 (1995), pp. 1353–1356. DOI: 10.1016/0005-1098(95)00042-U.

[2] R. H. Andersen, T. Solund, and J. Hallam. "Definition and Initial Case-Based Evaluation of Hardware-Independent Robot Skills for Industrial Robotic Co-Workers." In: *Proc. 41st Int. Symp. Robotics (ISR)*. Munich, Germany, 2014, pp. 1–7.

[3] F. Balduzzi, A. Giua, and G. Menga. "First-order hybrid Petri nets: a model for optimization and control." In: *IEEE Trans. Robotics and Automation* 16.4 (2000), pp. 382–399. DOI: 10.1109/70.864231.

[4] R. Beach, A. P. Muhlemann, D. H. R. Price, A. Paterson, and J. A. Sharp. "A review of manufacturing flexibility." In: *European J. Operational Research* 122.1 (2000), pp. 41–57. DOI: 10.1016/S0377-2217(99)00062-4.

[5] M. Bengel. "Model-based configuration - A workpiece-centred approach." In: *Proc. ASME/IFToMM Int. Conf. Reconfigurable Mechanisms and Robots (ReMar)*. London, United Kingdom, 2009, pp. 689–695.

[6] E. G. Birgin, P. Feofiloff, C. G. Fernandes, E. L. De Melo, M. T. I. Oshiro, and D. P. Ronconi. "A MILP model for an extended version of the flexible job shop problem." In: *Optimization Letters* 8.4 (2014), pp. 1417–1431. DOI: 10.1007/s11590-013-0669-7.

[7] E. G. Birgin, J. E. Ferreira, and D. P. Ronconi. "List scheduling and beam search methods for the flexible job shop scheduling problem with sequencing flexibility." In: *European J. Operational Research* 247.2 (2015), pp. 421–440. DOI: 10.1016/j.ejor.2015.06.023.

[8] B. Boss, S. Malakuti, S.-W. Lin, T. Usländer, E. Clauer, M. Hoffmeister, and L. Stojanovic. *Digital Twin and Asset Administration Shell Concepts and Application in the Industrial Internet and Industrie 4.0*. Tech. rep. Plattform Industrie 4.0 & Industrial Internet Consortium, 2020.

[9] P. Brandimarte. "Routing and scheduling in a flexible job shop by tabu search." In: *Annals of Operations Research* 41.3 (1993), pp. 157–183. DOI: 10.1007/BF02023073.

[10] P. Brucker. "Classification of Scheduling Problems." In: *Scheduling Algorithms.* Berlin/Heidelberg: Springer, 2007, pp. 1–10. DOI: `10.1007/978-3-540-69516-5_1`.

[11] C. G. Cassandras and S. Lafortune. *Introduction to Discrete Event Systems.* Berlin/Heidelberg, Germany: Springer Science & Business Media, 2009. DOI: `10.1007/978-0-387-68612-7`.

[12] A. Cataldo, A. Perizzato, and R. Scattolini. "Production scheduling of parallel machines with model predictive control." In: *Control Engineering Practice* 42 (2015), pp. 28–40. DOI: `10.1016/j.conengprac.2015.05.007`.

[13] I. A. Chaudhry and A. A. Khan. "A research survey: review of flexible job shop scheduling techniques." In: *Int. Trans. Operational Research* 23.3 (2016), pp. 551–591. DOI: `10.1111/itor.12199`.

[14] Y. Cohen, H. Naseraldin, A. Chaudhuri, and F. Pilati. "Assembly systems in Industry 4.0 era: a road map to understand Assembly 4.0." In: *The Int. J. Advanced Manufacturing Technology* 105.9 (2019), pp. 4037–4054. DOI: `10.1007/s00170-019-04203-1`.

[15] Y. Demir and S. K. İşleyen. "Evaluation of mathematical models for flexible job-shop scheduling problems." In: *Appl. Math. Modelling* 37.3 (2013), pp. 977–988. DOI: `10.1016/j.apm.2012.03.020`.

[16] M. Diaz. *Petri Nets.* Hoboken, NJ: John Wiley & Sons, Ltd, 2009. DOI: `10.1002/9780470611647`.

[17] C. Dripke, B. Schneider, M. Dragan, A. Zoitl, and A. Verl. "Concept of Distributed Interpolation for Skill-Based Manufacturing with Real-Time Communication." In: *Tagungsband des 3. Kongresses Montage Handhabung Industrieroboter.* Ed. by T. Schüppstuhl, K. Tracht, and J. Franke. Berlin/Heidelberg, Germany: Springer, 2018, pp. 215–222. DOI: `10.1007/978-3-662-56714-2_24`.

[18] S. A. Fahmy, T. Y. ElMekkawy, and S. Balakrishnan. "Analysis of reactive deadlock-free scheduling in flexible job shops." In: *Int. J. Flexible Manufacturing Systems* 19.3 (2007), pp. 264–285. DOI: `10.1007/s10696-007-9026-4`.

[19] T. Faulwasser, L. Grüne, and M. A. Müller. "Economic Nonlinear Model Predictive Control." In: *Foundations and Trends® in Systems and Control* 5.1 (2018), pp. 1–98. DOI: `10.1561/2600000014`.

[20] C. Feller and C. Ebenbauer. "Relaxed Logarithmic Barrier Function Based Model Predictive Control of Linear Systems." In: *IEEE Trans. Automat. Control* 62.3 (2017), pp. 1223–1238. DOI: `10.1109/TAC.2016.2582040`.

[21] C. Feller and C. Ebenbauer. "Sparsity-Exploiting Anytime Algorithms for Model Predictive Control: A Relaxed Barrier Approach." In: *IEEE Trans. Control Systems Technology* 28.2 (2018), pp. 425–435. DOI: `10.1109/TCST.2018.2880142`.

[22] M. G. Forbes, R. S. Patwardhan, H. Hamadah, and R. B. Gopaluni. "Model Predictive Control in Industry: Challenges and Opportunities." In: *Proc. 9th IFAC Symp. Advanced Control of Chemical Processes (ADCHEM)*. Whistler, Canada, 2015, pp. 531–538. DOI: `10.1016/j.ifacol.2015.09.022`.

[23] J. Friedrich, S. Scheifele, A. Verl, and A. Lechler. "Flexible and Modular Control and Manufacturing System." In: *Procedia CIRP* 33 (2015), pp. 115–120. DOI: `10.1016/j.procir.2015.06.022`.

[24] K. Genova, L. Kirilov, and V. Guliashki. "A Survey of Solving Approaches for Multiple Objective Flexible Job Shop Scheduling Problems." In: *Cybernetics and Information Technologies* 15.2 (2015), pp. 3–22. DOI: `10.1007/s10845-013-0837-8`.

[25] A. Giua and C. Seatzu. "A systems theory view of Petri nets." In: *Advances in control theory and applications*. Ed. by C. Bonivento, L. Marconi, C. Rossi, and A. Isidori. Berlin/Heidelberg, Germany: Springer, 2007, pp. 99–127. DOI: `10.1007/978-3-540-70701-1_6`.

[26] R. L. Graham, E. L. Lawler, J. K. Lenstra, and A. H. G. Rinnooy Kan. "Optimization and Approximation in Deterministic Sequencing and Scheduling: A Survey." In: *Discrete Optimization II*. Ed. by P. Hammer, E. Johnson, and B. Korte. Vol. 5. Annals of Discrete Mathematics. Elsevier, 1979, pp. 287–326. DOI: `10.1016/S0167-5060(08)70356-X`.

[27] M. Grieves. *Digital Twin: Manufacturing Excellence Through Virtual Factory Replication*. Tech. rep. Florida Institute of Technology, 2014.

[28] L. Grüne and J. Pannek. *Nonlinear Model Predictive Control: Theory and Algorithms*. Cham, Switzerland: Springer Int. Publishing, 2017.

[29] R. Heim, P. M. S. Nazari, J. O. Ringert, B. Rumpe, and A. Wortmann. "Modeling Robot and World Interfaces for Reusable Tasks." In: *Proc. IEEE/RSJ Int. Conf. Intelligent Robots and Systems (IROS)*. Hamburg, Germany, 2015, pp. 1793–1798. DOI: `10.1109/IROS.2015.7353610`.

[30] H. Herrero, A. A. Moughlbay, J. L. Outón, D. Sallé, and K. L. de Ipiña. "Skill based robot programming: Assembly, vision and Workspace Monitoring skill interaction." In: *Neurocomputing* 255 (2017), pp. 61–70. DOI: `10.1016/j.neucom.2016.09.133`.

[31] J. W. Herrmann. "A History of Production Scheduling." In: *Handbook of Production Scheduling*. Ed. by J. W. Herrmann. Boston, MA: Springer, 2006, pp. 1–22. DOI: `10.1007/0-387-33117-4_1`.

[32] L. Heuss, A. Blank, S. Dengler, G. L. Zikeli, G. Reinhart, and J. Franke. "Modular Robot Software Framework for the Intelligent and Flexible Composition of Its Skills." In: *Proc. Advances in Production Management Systems. Production Management for the Factory of the Future (APMS)*. Austin, TX, USA, 2019, pp. 248–256. DOI: `10.1007/978-3-030-30000-5_32`.

[33] W. K. Holstein and M. Tanenbaum. "production system." In: *Encyclopedia Britannica*. https://www.britannica.com/technology/production-system, 2020. Accessed 5 January 2022.

[34] M. ten Hompel, B. Vogel-Heuser, and T. Bauernhansl, eds. *Handbuch Industrie 4.0*. Berlin/Heidelberg, Germany: Springer Vieweg, 2016. DOI: 10.1007/978-3-662-45537-1.

[35] H. H. Hoos and T. Stützle. "Scheduling Problems." In: *Stochastic Local Search*. Ed. by H. H. Hoos and T. Stützle. Morgan Kaufmann, 2005. Chap. 9, pp. 417–465. DOI: 10.1016/B978-155860872-6/50026-3.

[36] F. Jaensch, A. Csiszar, C. Scheifele, and A. Verl. "Digital Twins of Manufacturing Systems as a Base for Machine Learning." In: *Proc. 25th Int. Conf. Mechatronics and Machine Vision in Practice (M2VIP)*. Stuttgart, Germany, 2018, pp. 1–6. DOI: 10.1109/M2VIP.2018.8600844.

[37] A. Jain, P. K. Jain, F. T. S. Chan, and S. Singh. "A review on manufacturing flexibility." In: *Int. J. Production Research* 51.19 (2013), pp. 5946–5970. DOI: 10.1080/00207543.2013.824627.

[38] J. Júlvez, S. Di Cairano, A. Bemporad, and C. Mahulea. "Event-driven model predictive control of timed hybrid Petri nets." In: *Int. J. Robust and Nonlinear Control* 24.12 (2014), pp. 1724–1742. DOI: 10.1002/rnc.2958.

[39] H. Kagermann, J. Helbig, A. Hellinger, and W. Wahlster. *Recommendations for implementing the strategic initiative INDUSTRIE 4.0: Securing the future of German manufacturing industry; final report of the Industrie 4.0 Working Group*. Munich, Germany: Forschungsunion, 2013.

[40] Y. K. Kim, K. Park, and J. Ko. "A symbiotic evolutionary algorithm for the integration of process planning and job shop scheduling." In: *Computers & Operations Research* 30.8 (2003), pp. 1151–1171. DOI: 10.1016/S0305-0548(02)00063-1.

[41] P. Köhler, M. A. Müller, J. Pannek, and F. Allgöwer. "On Exploitation of Supply Chain Properties by Sequential Distributed MPC." In: *Proc. 20th IFAC World Congress*. Toulouse, France, 2017, pp. 8219–8224. DOI: 10.1016/j.ifacol.2017.08.706.

[42] P. N. Köhler, M. A. Müller, and F. Allgöwer. "A distributed economic MPC framework for cooperative control under conflicting objectives." In: *Automatica* 96 (2018), pp. 368–379. DOI: 10.1016/j.automatica.2018.07.001.

[43] G. M. Kopanos, E. Capon-Garcia, A. Espuna, and L. Puigjaner. "Costs for Rescheduling Actions: A Critical Issue for Reducing the Gap between Scheduling Theory and Practice." In: *Industrial & Engineering Chemistry Research* 47.22 (2008), pp. 8785–8795. DOI: 10.1021/ie8005676.

[44] F. Kretschmer. "Gelbe Seiten für Industrie 4.0." In: *Automat!on praxis* (2016).

[45] J.-I. Latorre-Biel, J. Faulín, A. A. Juan, and E. Jiménez-Macías. "Petri Net Model of a Smart Factory in the Frame of Industry 4.0." In: *Proc. 9th Vienna Int. Conf. Math. Modelling*. Vienna, Austria, 2018, pp. 266–271. DOI: 10.1016/j.ifacol.2018.03.046.

[46] D. Limon, T. Alamo, F. Salas, and E. Camacho. "On the stability of constrained MPC without terminal constraint." In: *IEEE Trans. Automat. Control* 51.5 (2006), pp. 832–836. DOI: 10.1109/TAC.2006.875014.

[47] R. Lindorfer and R. Froschauer. "Towards user-oriented programming of skill-based automation systems using a domain-specific meta-modeling approach." In: *Proc. IEEE 17th Int. Conf. Industrial Informatics (INDIN)*. Helsinki-Espoo, Finland, 2019, pp. 655–660. DOI: 10.1109/INDIN41052.2019.8972318.

[48] K. D. Listmann, P. Wenzelburger, and F. Allgöwer. "Industrie 4.0 - (R)evolution without Control Technologies?" In: *J. Society of Instrument and Control Engineers* 55.7 (2016), pp. 555–565. DOI: 10.11499/sicejl.55.555.

[49] F. Long, P. Zeiler, and B. Bertsche. "Modelling the flexibility of production systems in Industry 4.0 for analysing their productivity and availability with high-level Petri nets." In: *Proc. 20th IFAC World Congress*. Toulouse, France, 2017, pp. 5680–5687. DOI: 10.1016/j.ifacol.2017.08.1118.

[50] Z. Luan, J. Zhang, W. Zhao, and C. Wang. "Trajectory Tracking Control of Autonomous Vehicle With Random Network Delay." In: *IEEE Trans. Vehicular Technology* 69.8 (2020), pp. 8140–8150. DOI: 10.1109/TVT.2020.2995408.

[51] W. T. Lunardi, E. G. Birgin, P. Laborie, D. P. Ronconi, and H. Voos. "Mixed Integer linear programming and constraint programming models for the online printing shop scheduling problem." In: *Computers & Operations Research* 123 (2020), p. 105020. DOI: 10.1016/j.cor.2020.105020.

[52] J. M. Maestre and R. R. Negenborn, eds. *Distributed model predictive control made easy*. Vol. 69. Berlin/Heidelberg, Germany: Springer, 2014. DOI: 10.1007/978-94-007-7006-5.

[53] C. Mahulea, A. Giua, L. Recalde, C. Seatzu, and M. Silva. "Optimal model predictive control of timed continuous Petri nets." In: *IEEE Trans. Automatic Control* 53.7 (2008), pp. 1731–1735. DOI: 10.1109/TAC.2008.929386.

[54] S. Malakuti, J. Bock, M. Weser, P. Venet, P. Zimmermann, M. Wiegand, J. Grothoff, C. Wagner, and A. Bayha. "Challenges in Skill-based Engineering of Industrial Automation Systems." In: *Proc. IEEE 23rd Int. Conf. Emerging Technologies and Factory Automation (ETFA)*. Turin, Italy, 2018, pp. 67–74. DOI: 10.1109/ETFA.2018.8502635.

[55] A. S. Manne. "On the job-shop scheduling problem." In: *Operations Research* 8.2 (1960), pp. 219–223. DOI: 10.1287/opre.8.2.219.

[56] R. D. McAllister, J. B. Rawlings, and C. T. Maravelias. "Rescheduling Penalties for Economic Model Predictive Control and Closed-Loop Scheduling." In: *Industrial & Engineering Chemistry Research* 59.6 (2020), pp. 2214–2228. DOI: 10.1021/acs.iecr.9b05255.

[57] M. A. Müller and L. Grüne. "Economic model predictive control without terminal constraints for optimal periodic behavior." In: *Automatica* 70 (2016), pp. 128–139. DOI: 10.1016/j.automatica.2016.03.024.

[58] M. A. Müller and K. Worthmann. "Quadratic costs do not always work in MPC." In: *Automatica* 82 (2017), pp. 269–277. DOI: 10.1016/j.automatica.2017.04.058.

[59] T. Murata. "Petri Nets: Properties, Analysis and Applications." In: *Proceedings of the IEEE* 77.4 (1989), pp. 541–580. DOI: 10.1109/5.24143.

[60] M. M. Nasiri and F. Kianfar. "A hybrid scatter search for the partial job shop scheduling problem." In: *The Int. J. Advanced Manufacturing Technology* 52.9 (2011), pp. 1031–1038. DOI: 10.1007/s00170-010-2792-2.

[61] E. Negri, L. Fumagalli, and M. Macchi. "A review of the roles of digital twin in CPS-based production systems." In: *Procedia Manufacturing* 11 (2017), pp. 939–948. DOI: 10.1016/j.promfg.2017.07.198.

[62] L. Ollinger, J. Schlick, and S. Hodek. "Leveraging the Agility of Manufacturing Chains by Combining Process-Oriented Production Planning and Service-Oriented Manufacturing Automation." In: *Proc. 18th IFAC World Congress.* Milano, Italy, 2011, pp. 5231–5236. DOI: 10.3182/20110828-6-IT-1002.01834.

[63] C. Özgüven, L. Özbakır, and Y. Yavuz. "Mathematical models for job-shop scheduling problems with routing and process plan flexibility." In: *Appl. Math. Modelling* 34.6 (2010), pp. 1539–1548. DOI: 10.1016/j.apm.2009.09.002.

[64] C. A. Petri. "Kommunikation mit Automaten." PhD thesis. Technische Hochschule Darmstadt, 1962.

[65] M. L. Pinedo. "Deterministic Models: Preliminaries." In: *Scheduling: Theory, Algorithms, and Systems.* Cham, Switzerland: Springer Int. Publishing, 2016, pp. 13–32. DOI: 10.1007/978-3-319-26580-3_2.

[66] M. L. Pinedo. *Scheduling: Theory, Algorithms, and Systems.* 5th ed. Cham, Switzerland: Springer Int. Publishing, 2016. DOI: 10.1007/978-3-319-26580-3.

[67] J. B. Rawlings, D. Q. Mayne, and M. M. Diehl. *Model Predictive Control: Theory, Computation, and Design.* Madison, WI, USA: Nob Hill Publishing, 2017.

[68] L. Recalde, E. Teruel, and M. Silva. "Autonomous continuous P/T systems." In: *Proc. Int. Conf. Application and Theory of Petri Nets (ICATPN).* Williamsburg, VA, USA, 1999, pp. 107–126. DOI: 10.1007/3-540-48745-X_8.

[69] C. Reiff, M. Buser, T. Betten, V. Onuseit, M. Hoßfeld, D. Wehner, and O. Riedel. "A Process-Planning Framework for Sustainable Manufacturing." In: *Energies* 14.18 (2021), p. 5811. DOI: 10.3390/en14185811.

[70] D. A. Rossit, F. Tohmé, and M. Frutos. "Industry 4.0: Smart Scheduling." In: *Int. J. Production Research* 57.12 (2019), pp. 3802–3813. DOI: 10.1080/00207543. 2018.1504248.

[71] C. Schlegel, A. Lotz, M. Lutz, D. Stampfer, J. F. Inglés-Romero, and C. Vicente-Chicote. "Model-driven software systems engineering in robotics: covering the complete life-cycle of a robot." In: *it-Information Technology* 57.2 (2015), pp. 85–98. DOI: 10.1515/itit-2014-1069.

[72] P. Scokaert and J. Rawlings. "Constrained linear quadratic regulation." In: *IEEE Trans. Automat. Control* 43.8 (1998), pp. 1163–1169. DOI: 10.1109/9.704994.

[73] C. Seatzu, M. Silva, and J. H. Van Schuppen. *Control of discrete-event systems*. Vol. 433. Berlin/Heidelberg, Germany: Springer, 2013. DOI: 10.1007/978-1-4471-4276-8.

[74] A. K. Sethi and S. P. Sethi. "Flexibility in manufacturing: a survey." In: *Int. J. of Flexible Manufacturing Systems* 2.4 (1990), pp. 289–328. DOI: 10.1007/BF00186471.

[75] M. Silva and L. Recalde. "On fluidification of Petri Nets: from discrete to hybrid and continuous models." In: *Annual Reviews in Control* 28.2 (2004), pp. 253–266. DOI: 10.1016/j.arcontrol.2004.05.002.

[76] K. Subramanian, C. T. Maravelias, and J. B. Rawlings. "A state-space model for chemical production scheduling." In: *Comp. & Chem. Eng.* 47 (2012), pp. 97–110. DOI: 10.1016/j.compchemeng.2012.06.025.

[77] K. Subramanian, J. B. Rawlings, C. T. Maravelias, J. Flores-Cerrillo, and L. Megan. "Integration of control theory and scheduling methods for supply chain management." In: *Comp. & Chem. Eng.* 51 (2013), pp. 4–20. DOI: 10.1016/j. compchemeng.2012.06.012.

[78] M. Taleb, E. Leclercq, and D. Lefebvre. "Model predictive control for discrete and continuous timed Petri nets." In: *Int. J. Automation and Computing* 15.1 (2018), pp. 25–38. DOI: 10.1007/s11633-016-1046-7.

[79] U. Thomas, G. Hirzinger, B. Rumpe, C. Schulze, and A. Wortmann. "A New Skill Based Robot Programming Language Using UML/P Statecharts." In: *Proc. IEEE Int. Conf. Robotics and Automation (ICRA)*. Karlsruhe, Germany, 2013, pp. 461–466. DOI: 10.1109/ICRA.2013.6630615.

[80] A. Türkyılmaz, Ö. Şenvar, İ. Ünal, and S. Bulkan. "A research survey: heuristic approaches for solving multi objective flexible job shop problems." In: *J. Intelligent Manufacturing* 31.8 (2020), pp. 1949–1983. DOI: 10.1007/s10845-020-01547-4.

[81]  F. D. Vargas-Villamil and D. E. Rivera. "A model predictive control approach for real-time optimization of reentrant manufacturing lines." In: *Computers in Industry* 45.1 (2001), pp. 45–57. DOI: 10.1016/S0166-3615(01)00080-X.

[82]  G. Vilcot and J.-C. Billaut. "A tabu search and a genetic algorithm for solving a bicriteria general job shop scheduling problem." In: *European J. Operational Research* 190.2 (2008), pp. 398–411. DOI: 10.1016/j.ejor.2007.06.039.

[83]  M. Wächter, S. Ottenhaus, M. Kröhnert, N. Vahrenkamp, and T. Asfour. "The armarx statechart concept: Graphical programing of robot behavior." In: *Frontiers in Robotics and AI* 3 (2016), p. 33. DOI: 10.3389/frobt.2016.00033.

[84]  B. Waschneck, T. Altenmüller, T. Bauernhansl, and A. Kyek. "Production Scheduling in Complex Job Shops from an Industry 4.0 Perspective: A Review and Challenges in the Semiconductor Industry." In: *Proc. 1st Int. Workshop on Science, Application and Methods in Industry 4.0*. Graz, Austria, 2016, pp. 1–12.

[85]  B. Waschneck, A. Reichstaller, L. Belzner, T. Altenmüller, T. Bauernhansl, A. Knapp, and A. Kyek. "Optimization of global production scheduling with deep reinforcement learning." In: *Procedia Cirp* 72 (2018), pp. 1264–1269. DOI: 10.1016/j.procir.2018.03.212.

[86]  P. Wenzelburger and F. Allgöwer. "A Novel Optimal Online Scheduling Scheme for Flexible Manufacturing Systems." In: *Proc. 13th IFAC Workshop on Intelligent Manufacturing Systems (IMS)*. Oshawa, Canada, 2019, pp. 1–6. DOI: 10.1016/j.ifacol.2019.10.002.

[87]  P. Wenzelburger and F. Allgöwer. "A Petri Net Modeling Framework for the Control of Flexible Manufacturing Systems." In: *Proc. 9th IFAC Conf. Manufacturing Modeling, Management, and Control (MIM)*. Berlin, Germany, 2019, pp. 492–498. DOI: 10.1016/j.ifacol.2019.11.111.

[88]  P. Wenzelburger and F. Allgöwer. "Model Predictive Control for Flexible Job Shop Scheduling in Industry 4.0." In: *Applied Sciences* 11.17 (2021), p. 8145. DOI: 10.3390/app11178145.

[89]  W. Xiang and H. P. Lee. "Ant colony intelligence in multi-agent dynamic manufacturing scheduling." In: *Engineering Applications of Artificial Intelligence* 21.1 (2008), pp. 73–85. DOI: 10.1016/j.engappai.2007.03.008.

[90]  A. Yadav and S. C. Jayswal. "Modelling of flexible manufacturing system: a review." In: *Int. J. Production Research* 56.7 (2018), pp. 2464–2487. DOI: 10.1080/00207543.2017.1387302.

[91]  J. Zhang, G. Ding, Y. Zou, S. Qin, and J. Fu. "Review of job shop scheduling research and its new perspectives under Industry 4.0." In: *J. Intelligent Manufacturing* 30.4 (2019), pp. 1809–1830. DOI: 10.1007/s10845-017-1350-2.

[92] Q. Zhang, P. Liu, and J. Pannek. "Modeling and predictive capacity adjustment for job shop systems with RMTs." In: *Proc. 25th Mediterranean Conf. Control and Automation (MED)*. Valletta, Malta, 2017, pp. 310–315. DOI: 10.1109/MED.2017.7984136.

[93] R. Y. Zhong, X. Xu, E. Klotz, and S. T. Newman. "Intelligent Manufacturing in the Context of Industry 4.0: A Review." In: *Engineering* 3.5 (2017), pp. 616–630. DOI: 10.1016/J.ENG.2017.05.015.

[94] T. K. Zubaran and M. Ritt. "An effective heuristic algorithm for the partial shop scheduling problem." In: *Computers & Operations Research* 93 (2018), pp. 51–65. DOI: 10.1016/j.cor.2018.01.015.